97-HHE-1

Development of Practical Methods to Assess the Presence of Bacterial Pathogens in Water

by:

William A. Yanko
County Sanitation Districts of Los Angeles County (Retired)

Ricardo De Leon, Ph.D.
Metropolitan Water District of Southern California

Paul A. Rochelle, Ph.D.
Metropolitan Water District of Southern California

Wilfred Chen, Ph.D.
University of California, Riverside

2004

Water Environment Research Foundation

The Water Environment Research Foundation, a not-for-profit organization, funds and manages water quality research for its subscribers through a diverse public-private partnership between municipal utilities, corporations, academia, industry, and the federal government. WERF subscribers include municipal and regional water and wastewater utilities, industrial corporations, environmental engineering firms, and others that share a commitment to cost-effective water quality solutions. WERF is dedicated to advancing science and technology addressing water quality issues as they impact water resources, the atmosphere, the lands, and quality of life.

For more information, contact:
Water Environment Research Foundation
635 Slaters Lane, Suite 300
Alexandria, VA 22314-1177
Tel: (703) 684-2470
Fax: (703) 299-0742
www.werf.org
werf@werf.org

This report was co-published by the following organization. For nonsubscriber sales information, contact:

IWA Publishing
Alliance House, 12 Caxton Street
London SW1H 0QS, United Kingdom
Tel: +44 (0) 20 7654 5500
Fax: +44 (0) 20 7654 5555
www.iwapublishing.com
publications@iwap.co.uk

Library of Congress Catalog Card Number: 2003096167

IWAP ISBN: 1-84339-688-2

County Sanitation Districts of Los Angeles County, Metropolitan Water District of Southern California, University of California, Riverside

The research on which this report is based was funded, in part, by the United States Environmental Protection Agency through Cooperative Agreement No. CR-825237-01 with the Water Environment Research Foundation (WERF). Unless an EPA logo appears on the cover, this report is a publication of WERF, not EPA. Funds awarded under the Cooperative Agreement cited above were not used for editorial services, reproduction, printing, or distribution.

ACKNOWLEDGMENTS

The authors wish to express their appreciation to the Project Subcommittee members for their helpful suggestions and patience during this study. We also wish to thank the numerous agencies and personnel in California, New York, Ohio, Texas, and Washington who offered to assist and provide samples for this project.

Report Preparation

Principal Investigators:
William A. Yanko
County Sanitation Districts of Los Angeles County

Ricardo DeLeon, Ph.D.
Paul A. Rochelle, Ph.D.
Metropolitan Water District of Southern California

Wilfred Chen, Ph.D.
University of California, Riverside

Project Team:
Alan Walker
Jennipher Cu
April Garcia
Letty Libao
Julie Millenbach
County Sanitation Districts of Los Angeles County

Patrick Hacker
Anagha Chitre
Katrin Hanley
Jason La
Patty Huang
Anne M. Johnson
Metropolitan Water District of Southern California

Ashok Mulchandani, Ph.D.
Nathalie Fortin, Ph.D.
Grisselle Martinez, M.S.
University of California, Riverside

Project Subcommittee

Jay Vasconcelos, Ph.D., Chair
Watertrek Consulting Services, Inc.

Stephanie I. Harris, DVM
United States Environmental Protection Agency

Gregory J. Ma, SM (AAM), MSPH
King County Environmental Laboratory

Pierre Payment, Ph.D.
INRS-Institut Armand-Frappier

Lynne Putnam
Black & Veatch

Joyce M. Perdek, P.E.
United States Environmental Protection Agency

Izabela Wojtenko, Ph.D.
United States Environmental Protection Agency

Water Environment Research Foundation Staff

Director of Research:	Daniel M. Woltering, Ph.D.
Project Manager:	Bonnie Bailey

ABSTRACT

The objective of this project was to develop and evaluate a practical technique to indicate the probability of the presence of bacterial pathogens in receiving waters. A "practical" method was defined as one that would limit the use of defined culture-based microbiological methods, and would be based on: 1) validating indicator organisms that predicted the presence of pathogens or 2) detection of pathogens by polymerase chain reaction (PCR)-based assays. The study also assessed the utility of PCR-based technology for bacterial pathogen detection with respect to technology transfer to a wider range of water and wastewater facilities.

A simple, low cost, method to concentrate 10-L volumes of water was developed and validated. The method is adaptable for use by smaller utilities and would enable local personnel to collect large volume samples and reduce the volume to a size suitable for shipping to a reference or contract laboratory for bacterial pathogen analysis. The large-volume method, combined with overnight enrichment and PCR-based detection, was demonstrated to be a sensitive method for detecting selected aerobic bacterial pathogens within 24 hours with a sensitivity of 1 CFU/L. Two hundred samples, with water quality characteristics ranging from pristine to very poor, were analyzed using the enrichment/PCR approach: 90% were positive for *Aeromonas hydrophila*, 72% were positive for *Salmonella* spp., 34% for toxigenic coliforms, and 1.5% for *Yersinia enterocolitica*. Using an alternative direct PCR detection approach, the frequency of detection ranged from 0% to 55% for a wider range of pathogens, including 11% that were positive for *Mycobacterium avium*. A molecular beacon PCR assay, which combines amplification and confirmatory probing into a single reaction, was developed for detection of *Salmonella* spp. and *Escherichia coli* O157:H7. One urban stream sample was confirmed as positive for *E. coli* O157:H7 by two independent PCR-based analyses. Although the sample processing, concentration, and extraction procedures that precede molecular detection assays need further development before the methods can be adopted for routine monitoring purposes, molecular-based methods hold great promise for specific pathogen detection.

A relationship was demonstrated between the concentration of selected indicators and the probability of detecting 1) *Salmonella* spp. by culture and 2) shiga toxin-producing coliforms (*E. coli* O157 and related bacteria) by PCR. The approach of relating the concentration of indicator organisms to the probability of pathogens being present was shown to be a useful method to evaluate the predictive value of indicators. However, a larger database on the occurrence of other bacterial pathogens would be required to evaluate the predictive value of indicators for pathogens other than *Salmonella* and toxin gene-carrying coliforms.

BENEFITS

- A simple, low cost method for concentrating 10-L volumes of water for analysis of bacterial pathogens was validated.
- Integrated cultural and molecular techniques that can screen for an array of pathogens in environmental waters in a single sample were developed and validated. Potential applications include pathogen monitoring programs, source water protection, and screening of effluents.
- A real-time PCR assay for the detection of *Salmonella* spp. and *E. coli* O157:H7 based on the use of molecular beacons was developed, eliminating the need for secondary confirmation of the PCR products.
- The approach of correlating the concentration of microbial indicators to the probability of the presence of pathogens was demonstrated to be a useful method to evaluate the predictive value of indicators.
- The probability of detecting *Salmonella* spp. and shiga toxin-producing coliforms (STEC and related bacteria) clearly increased as the concentration of the *E. coli* and enterococci indicators increased, confirming a relationship between concentration of indicators and the probability of the presence of some enteric bacterial pathogens.
- The transfer of molecular-based technology from experienced to less experienced laboratories was demonstrated to be feasible. Most laboratories with trained microbiologists could be adapted to include some level of molecular-based testing without excessive cost or extensive training. Adequate quality assurance is a critical consideration.
- The water/wastewater industries can use the products and results of this project to assist in the decision making process related to the introduction and implementation of molecular methods into pathogen monitoring and source water protection programs.

Keywords: contamination, detection, filtration, concentration, microbial indicators, molecular, polymerase chain reaction, PCR.

TABLE OF CONTENTS

LIST OF TABLES

LIST OF FIGURES

LIST OF ACRONYMS

cDNA	Complementary DNA
CSDLAC	County Sanitation Districts of Los Angeles County
CFU	Colony forming unit
DE	Diatomaceous earth
DNA	Deoxyribonucleic acid
EDTA	Ethylenediaminetetraacetic acid
FRNA	RNA bacteriophage that infect the pilus of male (F) *E. coli* strains
GF/F-DE	Whatman glass fiber filter grade F overlaid with diatomaceous earth
MA	*Mycobacterium avium*
MAC	*Mycobacterium avium* complex
MI	*Mycobacterium intracelullarae*
MPN	Most probable number
mRNA	Messenger RNA
MWDSC	Metropolitan Water District of Southern California
PCR	Polymerase chain reaction
rRNA	Ribosomal RNA
RNA	Ribonucleic acid
STEC	Shiga toxin-producing strains of *E. coli*

EXECUTIVE SUMMARY

ES.1 Background

The primary goal of this project was to develop a practical monitoring technique that would indicate the probability of the presence of bacterial pathogens in receiving waters. The research plan assumed that a "practical" method for detecting bacterial pathogens would not utilize conventional culture-based microbiological analyses for pathogens. The proposal was based on the premise that a practical method would most likely be developed by 1) validating an indicator organism that provided meaningful predictive value for the presence of one or more bacterial pathogens or 2) on the direct detection of pathogens by molecular assay (PCR).

The research plan incorporated two parallel pathways. In one, the utility of the molecular techniques for the direct detection of the selected pathogens in surface water samples was examined. In the other pathway, a number of candidate indicator organisms were analyzed to examine the relationship between the concentrations of the indicators and the probability of the pathogen(s) being detected by molecular assay, or in selected cases by conventional culture assay.

The indicator organisms evaluated included total and fecal coliform and *E. coli*, fecal streptococci and enterococci, *Clostridium perfringens*, and somatic and male-specific coliphage. The pathogens selected for analysis by PCR were *Salmonella, Shigella, Campylobacter jejuni, Yersinia enterocolitica, E. coli* O157:H7, *Helicobacter pylori*, *Aeromonas hydrophila* and *Mycobacterium avian* complex (MAC). A subgroup of the pathogens, *Salmonella*, *E. coli* O157:H7 and *Aeromonas*, were also analyzed by conventional culture methods as a cross check for the PCR assays.

At the beginning of the study there was not a consensus that PCR-based assays would meet the definition of "practical" from the perspective of a routine water laboratory. For that reason the final portion of the study examined the potential for technology transfer and opportunities to make molecular assay technology more generally available for the water/wastewater industry.

ES.2 Method Evaluation/Validation

Detecting pathogens in water samples typically involves two main components, 1) concentration of the sample and 2) detection of the target pathogen(s) in the resulting concentrate. The first phase of this study was devoted to developing and evaluating the proposed methodology, including the pathogen analytical methods and a simple, low-cost method to concentrate large volumes of water for detecting bacterial pathogens. One of the difficulties in analyzing for waterborne pathogens such as bacteria is their intermittent presence in low numbers and their distribution in large volumes of surface waters. Therefore, sample concentration was considered a critical component of the study. A 10-L sample volume with a detection sensitivity of <10 cells per 10 L was selected as the target for sample concentration.

Methods must be low in initial cost and in operating cost to be successful for broad-based, application in the water/wastewater industry. The concentration method selected for 10-L samples used pressure filtration with inexpensive glass fiber filters (Whatman GF/F) overlaid with diatomaceous earth to increase bacterial capture and extend filtration volume. Validation studies demonstrated average retention of target pathogenic bacteria greater than 99%. Using culture-based assays, one viable bacterial cell per 10-L sample volume was detected 67% of the time, and the goal of detecting <10 cells per 10-L was met 88% of the time. The concentration method is readily adaptable for use by smaller utilities; where local resources cannot support pathogen analyses, large volume samples could be collected for pathogens and reduced in volume to a size suitable for shipping to a reference or contract laboratory for analysis.

The initial study plan called for direct PCR analysis of all the selected pathogens in the 10-L concentrates and for analysis of selected pathogens using culture methods. The 10-L concentrates were readily amenable to analysis using culture methods; however, culture-based methods are often complex and take many days to complete. For many pathogens, culture-based methods are insensitive and unable to detect pathogens potentially present. The 10-L concentrates were unfortunately found to be incompatible for direct molecular analyses, due to the presence of interfering substances that were also concentrated with the samples. Available methods to treat the concentrates were cumbersome, inefficient, and time consuming. To alleviate this problem, the concentrates were placed in a nonselective enrichment broth (Trypticase soy broth) and incubated overnight to amplify a wide range of aerobic pathogens prior to DNA extraction and PCR analysis. The overnight enrichment procedure solved the interference problem, increased detection sensitivity, and increased the probability that viable pathogens were being detected by PCR. The approach could also be used with selective enrichment to better target specific pathogens. The chief drawbacks to the addition of the nonselective enrichment step for molecular analysis were that it added 12 to 18 hours to the minimum time required from sample collection to reporting results, and the enrichment broth was not suitable for those pathogens with fastidious growth requirements.

The question of what constituted an appropriate volume to analyze was an important consideration during the formulation of the original study plan. The PCR technique might potentially increase detection sensitivity to the point that large volume samples were unnecessary. To address this possibility, a second, simpler concentration strategy was employed and validated during Phase 1. The membrane dissolution method consisted of filtering 1 L of sample through a 47-mm polycarbonate filter, then dissolving the filter in a solvent and extracting the DNA for PCR analysis. The procedure was straightforward and simple, and during the Phase 1 evaluation it was found suitable for clean water samples. Turbid samples required the use of more than one filter. However, the method proved to be more problematic than anticipated during the second phase of the study. With the range of surface water quality encountered, it was rarely possible to filter 1 L of sample, even when using two to three filters.

PCR primers for the selected pathogens were also evaluated during Phase 1 as part of the process to develop and validate the analytical strategy. PCR conditions were optimized and specificity of the primers verified by testing against a series of 44 other bacteria found in environmental water samples. Probe confirmation is another important step in the application of molecular techniques for detection of organisms in environmental samples. Technology transfer protocols, as well as quality control programs, should require the incorporation of a probe confirmation step. This step can be achieved via conventional membrane hybridization of Southern transfers or directly blotted amplicons on a membrane, hybridization in a microplate format, by molecular beacons during PCR, or by microarray technology. In this project, a probe confirmation step was conducted and optimized for each target amplicon by conventional membrane hybridization and for a selected set of organisms by molecular beacons. Oligonucleotides specific for target organism PCR products were designed and synthesized to confirm positive PCR results by probe hybridization as well as potentially increasing detection sensitivity.

As a further development of PCR-based detection technology, molecular beacon-PCR assays were developed for detection of *Salmonella* spp. and *Escherichia coli* O157:H7. This assay format combined amplification and confirmatory probing into a single homogeneous reaction and greatly reduced the analysis time compared to the conventional PCR/probe method. At least one urban stream sample was confirmed as positive for *E. coli* O157:H7 by both conventional PCR/probing and the molecular beacon-PCR assay.

ES.3 Monitoring Program

The second phase of the project was an extensive monitoring program to field test the selected methods and develop a database for the occurrence of pathogenic bacteria and concentrations of the indicator organisms. The majority of the 200 samples analyzed were collected in California in order to utilize existing sampling programs operated by the project agencies; however, 10% of the samples were collected in other regions of the U.S. Sample locations were categorized as: 1) untreated surface water, 2) urban runoff-influenced streams, or 3) wastewater effluents. Sampling locations included sites throughout the extensive watershed area that provides water supply to Southern California. In addition, local streams in the Los Angeles area that are not associated with water supply, but are heavily affected by wastewater effluents and urban runoff, were sampled. Sixteen (8%) of the samples were wastewater effluents representing different treatment plant sizes and levels of treatment. Other regions of the U.S. sampled included sites in New York, Ohio, Texas, and Washington. The sampling program represented a broad range in surface water quality and included seasonal variation.

A large database on the relative concentrations of the selected indicator organisms was developed and is tabulated in the report appendices. Geometric mean concentrations for the various indicators ranged from 20 to 22,000 per 100 mL. In surface water samples, there was a notable difference in the concentration of indicator organisms in the samples categorized as untreated surface water compared to those categorized as urban runoff-influenced streams, with the urban runoff-influenced streams having higher concentrations of all indicator organisms. Disinfected effluents had the lowest concentrations of the conventional indicators, coliforms and streptococci. Concentrations of *Clostridium perfringens* and coliphage, both somatic and male specific, were higher in disinfected effluents compared to most surface water samples. Excluding the disinfected samples, all the indicator groups tested in surface water during this study exhibited a strong positive correlation with each other (Spearman Rank Order Correlation Test). There was no indication that any of the indicators in surface waters trend independently.

When combined with PCR detection, the 10-L GF/F filtration method was an efficient, practical method for detecting aerobic bacterial pathogens. *Aeromonas hydrophila, Salmonella* spp. and shiga toxin-producing coliforms were detected in 90%, 72%, and 34% of the samples respectively. In comparison, *Aeromonas* and *Salmonella* spp. were detected in 96% and 61% respectively of the sample subset analyzed by culture methods, and no *E. coli* O157:H7 were detected by culture. However, the culture test was specific for *E. coli* O157:H7, whereas the molecular test detected any bacterial strain possessing the shiga toxin gene.

Detection frequency of the aerobic pathogens using the 47 mm polycarbonate membrane dissolution/direct PCR assay was significantly lower compared to the 10-L enrichment/PCR method and/or culture methods for those organisms tested by both methods, respectively. Due to the sampling processing difficulties, effectively less than 1 mL of sample was actually analyzed by an individual PCR reaction. In spite of the relatively low detection frequency using the membrane dissolution method, one pathogen having fastidious growth requirements, *Mycobacterium avium*, was detected in 12% of the samples. In addition, *Aeromonas hydrophila* was detected in 55% of the samples. Both of these potential pathogens are ubiquitous in water. These data confirm the need for an effective concentration procedure to detect most of the pathogens of concern in water.

A primary goal of the project was to examine the relationship between microbial indicator concentrations and the probability of detecting pathogenic bacteria in water samples. Samples that were culture-positive for *Salmonella* were compared to the subset of samples that were culture-negative for *Salmonella*. Mean indicator organism populations were higher in the *Salmonella* culture-positive sample set; however, individual samples were sometimes very low, or negative for indicators when *Salmonella* was detected. A single sample providing a low or negative indicator count did not assure the absence of a pathogen. Positive relationships were demonstrated between the concentration of selected indicators and the probability of detecting some of the pathogens in the study. The probability of detecting *Salmonella* spp. by cultural methods clearly increased as the concentration of the indicator *E. coli* increased. Increasing concentrations of the indicators *E. coli* and enterococci were also good predictors of the presence of shiga toxin-producing coliforms (*E. coli* O157:H7 and related bacteria) based on PCR-derived detection. The results demonstrated that the approach of relating the concentration of indicator organisms to the probability of pathogens being present was a useful method to evaluate the predictive value of indicators. However, a larger database on the occurrence of other bacterial pathogens would be required to evaluate the predictive value of indicators for pathogens other than *Salmonella* and enterotoxigenic coliforms.

ES.4 Technology Transfer

This project originated with the premise that rapid and simple methods are needed for the collection, concentration, isolation/extraction, and identification of pathogenic bacteria in untreated surface water, finished water, and wastewater. For microbial monitoring techniques to be practical for the water and wastewater industries they must be simple and robust enough to be performed by utility microbiology laboratories on a routine basis. Such laboratories may have only limited experience of molecular methods, if they have any experience at all, and will have neither the personnel nor resources to pursue full scale testing and evaluation of new procedures. Therefore, one of the original objectives of this project was to demonstrate that the various sample concentration and molecular detection technologies could be transferred to previously inexperienced laboratories with a minimum of training. A self-contained portable PCR laboratory originally proposed for this potion of the study was unavailable from the manufacturer and could not be evaluated. Therefore, the technology transfer focused on the ability to transfer the selected molecular methods from a highly experienced molecular laboratory research setting to a new, dedicated, but much less experienced, molecular lab, with an emphasis on the ability of the two laboratories to obtain consistent results on split samples. There was generally good concordance between the results obtained at the two laboratories following technology transfer; however, there were some differences. There was agreement between the two laboratories for 86.7% of the split samples (combining all water samples and all pathogen targets, n = 60) when results were based on PCR detection alone. The level of agreement was reduced slightly to 73.3% based on the combined PCR/probing results, with most of the differences being accounted for by samples that yielded positive results at the experienced lab but negative at the less experienced lab, primarily due to a problem obtaining a detectable hybridization signal with the *A. hydrophila* probe. None of the extraction blanks processed by either laboratory during this phase of the project generated false-positive results with any of the five primer sets used, indicating the adequacy of QA/QC procedures by both the experienced and inexperienced laboratories during DNA extraction and PCR processes. Despite the minor differences in pathogen detection during this phase of the project, the results of technology transfer demonstrated the transferability and robustness of the procedures.

This experience suggests that, with the use of a commercially available self-contained PCR laboratory and trained microbiologists on staff, some level of molecular testing could be transferred to a less experienced laboratory without excessive cost or extensive training. Development of adequate QA/QC protocols would be a critical consideration.

ES.5 Conclusions

1. Methods were developed and validated for the detection of a range of enteric bacterial pathogens in environmental waters. These methods integrated cultural and molecular techniques and can be used to screen for an array of pathogens in a single sample. Applications for these methods include pathogen monitoring programs, source water protection, and screening of effluents.

2. The presence of *A. hydrophila*, *Salmonella* spp., STEC (*E. coli* or other enterotoxigenic coliforms), *Y. enterocolitica*, and *M. avium* was confirmed in source and receiving waters.

3. The approach of correlating the concentration of microbial indicators to the probability of the presence of pathogens was a useful method to evaluate the predictive value of indicators.

4. The probability of detecting *Salmonella* spp. by cultural methods and shiga toxin-producing coliforms (STEC and related bacteria) by molecular methods clearly increased as the concentration of the *E. coli* and enterococci indicators increased, confirming a relationship between concentration of indicators and the probability of the presence of some enteric bacterial pathogens.

5. The transfer of molecular-based technology from experienced to less experienced laboratories was demonstrated to be feasible. Most laboratories with trained microbiologists could be adapted to include some levels of molecular-based testing without excessive cost or extensive training. Quality assurance is a critical consideration.

6. The water/wastewater industries can use the products and results of this project to assist in the decision-making process related to the introduction and implementation of molecular methods into pathogen monitoring and source water protection programs. While it is unlikely that currently used cultural methods will be replaced by molecular methods en masse, the availability of such methods increases the range of questions that can be answered regarding microbial water quality.

CHAPTER 1.0

INTRODUCTION

1.1 Background and Rationale

Coliform bacteria have traditionally been used as an indicator for the potential presence of pathogens in water. Indicator organisms are used because the pathogens are usually present in low numbers and are difficult to detect. Assuring the absence of coliforms in potable water or treated wastewater has, for the most part, helped assure safe potable water supplies and minimize risks from wastewater effluents, especially as related to the presence of bacterial pathogens. However, the utility of the coliform indicator system to assess risks from protozoan and viral pathogens has been questioned and remains a topic of research and concern in the water industry. A recent study by the United States Environmental Protection Agency (U.S. EPA) indicated that coliform bacteria were usually detected in association with identified disease outbreaks but that routine coliform monitoring seldom predicted the potential for an outbreak to occur (Craun et al., 1997). It was suggested over 25 years ago that the total coliform group is not a reliable indicator of pollutants or the condition of a water source (Davis et al., 1977).

In addition to monitoring treated water, coliform concentrations in surface waters are commonly used as a measure of relative risk. Federal and state standards based on acceptable levels of total and/or fecal coliforms have been established for recreational use and shellfishing. Epidemiological research, however, has questioned the value of coliform concentrations to predict recreational risks from viral pathogens (Cabelli, 1982), and there is little data specifically relating coliform concentrations to the probability of the presence of bacterial pathogens in surface water. The absence of coliform bacteria in surface water may reflect a low probability of many human enteric bacterial pathogens being present but, given the many diverse sources of coliform bacteria, the relationship between specific numbers of coliform bacteria and human pathogenic bacteria is tenuous at best. The capacity of many bacteria, including human pathogens, to enter a viable but nonculturable state further weakens the utility of coliforms for evaluating water quality (Colwell and Grimes, 2000). Fecal coliforms are also widely recognized as being inadequate indicators for predicting the response of some pathogens (bacterial, viral, or protozoal) to disinfectants. Alternative indicator systems have been suggested but few have been directly correlated with the presence of bacterial pathogens. Among the proposed indicators are *Escherichia coli,* enterococci, *Clostridium perfringens*, and male-specific coliphage.

Despite the continuing debate over their actual predictive utility, the use of indicators remains an attractive option because concentration procedures are typically not necessary and the test procedures are usually much easier to perform than those for specific human pathogens, requiring less specialized equipment and reducing the potential hazards presented by the routine manipulation of pathogens.

Ideally, direct measurement of the pathogens in water would provide the best assessment of microbiological water quality. There are numerous impediments to this approach related to the difficulty of detecting and measuring pathogens in water, and generally only the largest water utilities with specialized laboratories perform routine analyses for pathogenic microorganisms. Newer molecular techniques, such as the polymerase chain reaction (PCR), may provide improved means for determining the presence of pathogenic bacteria on a routine and cost-effective basis, or provide the ability to directly relate the presence of an indicator organism to the presence of the pathogens.

1.1.1 Pathogenic Bacteria in Surface Waters

Much attention has been focused on the fate and occurrence of protozoan and viral pathogens in water but bacterial pathogens in water continue to represent a significant threat to public health. Some are relatively ubiquitous in the water environment, such as *Legionella* spp. and *Salmonella* spp. (Fliermans et al., 1981; Cherry, et al., 1972), while little is known about the relative occurrence of many others in surface water. Bacterial pathogens of potential concern in surface water include, but are not limited to, *Salmonella* spp., *Shigella* spp., *Campylobacter* spp., enterotoxigenic *E. coli*, *Yersinia enterocolitica*, *Helicobacter pylori,* and the *Mycobacterium avium* complex (MAC). Also, many opportunistic pathogenic bacteria occur naturally in water including species of *Legionella, Pseudomonas, Serratia, Acinetobacter, Chromobacterium, Achromobacter* and *Aeromonas*, while others may be washed into water from their natural habitat in soil or on vegetative matter such as *Bacillus* spp., *Enterobacter* spp., *Klebsiella* spp., *Actinomyces* spp., and *Streptomyces* spp.

Salmonella spp. and *Campylobacter* spp. are the most common bacterial agents associated with acute gastrointestinal infections in the United States, and contaminated food remains the primary means of transmission. Numerous studies have been conducted on the occurrence of *Salmonella* spp. in both fresh and marine environments but results have been highly variable due to variations in the sensitivity of accepted isolation procedures (Morinigo, 1993). There are over 2,300 recognized serotypes of *Salmonella* spp. and no single procedure is able to detect all of them. Cherry et al. (1972) detected *Salmonella* spp. in both pristine and polluted surface water, but the occurrence was higher in highly polluted water. Morinigo et al. (1990) reported salmonellae widely distributed and detected *Salmonella* spp. in the absence of usual indicators. Recent studies in Southern California also found *Salmonella* spp. widely distributed in surface water (Kinde et al., 1997).

Shigella spp. are also agents of acute gastrointestinal disease in humans. Shigellae invade the intestinal mucosa, producing dysentery characterized by abdominal pain, fever, and diarrhea. Outbreaks are usually associated with food and less frequently with water, although shigellae can be isolated in water and wastewater (Lindell and Quinn, 1973; Cody and Tischer, 1965). Outbreaks of shigellosis have also been associated with swimming (Makintubee et al., 1987; Rosenberg et al., 1976).

E. coli is a normal inhabitant of the human intestinal tract; however some *E. coli* serotypes (*E. coli* O157:H7) have been found to cause diarrheal disease in humans. These pathogenic *E. coli* are classified into five categories: enterotoxigenic, enterohemorrhagic, enteroinvasive, enteropathogenic and enteradhercnt-aggregative (Vial et al., 1988). Definitive identification requires the determination of the characteristic virulence factors associated with each group. Pathogenic *E. coli* are most often implicated in foodborne illness; however, several major waterborne outbreaks have been reported (Feng, 1995). A recent waterborne outbreak in Canada caused primarily by *E. coli* O157:H7 and *Campylobacter jejuni* resulted in over 2,300 cases of illness and seven deaths (O'Conner, 2002).

Y. enterocolitica is a pathogen of potential importance in water. *Y. enterocolitica* has become recognized worldwide as an important human pathogen and in several countries it is nearly as common as *Salmonella* and *Campylobacter* as a leading cause of chronic enteritis (Fenwick and McCarty, 1995). *Y. enterocolitica* is usually associated with sporadic cases of gastroenteritis in the U.S.; however, epidemiological investigations suggest the predominant pathogenic serotype isolated in the U.S. has been changing (Bissett et al., 1990). *Y. enterocolitica* serogroup 0:3 has become the most common species recovered from patients, reflecting the same pattern seen in other parts of the world suggesting this organism could potentially be an emerging pathogen in the U.S.

Helicobacter pylori has been recognized as an etiologic agent of type B chronic gastritis. Infection rates in adults range from approximately 40% in developed countries to 80% in developing countries, suggesting *H. pylori* is probably one of the most common bacterial infections of humans (Cover and Blaser, 1995). The full impact of this emerging pathogen on public health may still be underestimated. Although fecal-oral and oral-oral routes of transmission have been suggested (Klein et al., 1991), the mode of transmission of *H. pylori* has yet to be determined, though a waterborne route of transmission has been implicated. Klein et al. (1991) observed that Peruvian children who lived in homes with external water sources were approximately three times more likely to have *H. pylori* infection than were those whose homes had an internal water supply. The possibility of waterborne *H. pylori* transmission raises the need to determine the prevalence and fate of this organism in water. Several authors have demonstrated that *H. pylori* can survive in water in a viable state for many days and as a viable but nonculturable coccoid form for up to a year. Benaissa et al. (1995) demonstrated that the coccoid form of *H. pylori* can survive for up to seven days in water and is still viable and able to regrow into the bacillary form when ingested. At 7°C, *H. pylori* remained culturable when suspended in distilled water for 11-14 days, and in physiological saline for 16 days (West et al., 1990). Since *H. pylori* can survive in water, it is essential to determine its frequency of occurrence and concentration in wastewater, surface and receiving water, and treated drinking water. The isolation of *H. pylori* by conventional methods involves fastidious growth requirements and a lengthy three-day incubation period. Therefore, nonconventional methods such as PCR-based detection and probe hybridization may be necessary for an assessment of its prevalence in the environment.

Mycobacterium avium complex (MAC), an opportunistic human pathogen, is the most common disseminated bacterial infection in AIDS patients (Glover et al., 1995; Armstrong et al., 1985). It is second only to the AIDS wasting syndrome as the most frequent cause of death in these patients (Hardy et al., 1992). As many as 40% of patients with advanced AIDS may develop this infection (Horsburgh, 1991). Most patients with disseminated MAC (*M. avium* [MA], *M. intracellulare* [MI], and MX) disease are infected with MA. MAC may also be invasive in other immunocompromised conditions such as neoplastic disorders, combined immunodeficiency disease, post-heart transplant (Novick et al., 1990), and in patients with no obvious predisposing factors (Prince et al., 1991). The environment is considered the most likely source of MAC infection and these organisms have been recovered from water (Aronson et al., 1993; Collins and Yates, 1984). The harsh decontamination conditions employed during the selective isolation of mycobacteria (Glover et al., 1994) and the ability of some bacteria to exist in a viable but nonculturable state (Roszak and Colwell, 1987) suggest that the prevalence of MAC bacteria determined by conventional isolation methods may be an underestimation of the actual number.

Aeromonas spp. are important waterborne opportunistic pathogens of both humans and animals. They are more numerous than total coliforms in some fresh waters (Ashbolt et al., 1993), are frequently detected in distribution systems (Payment et al., 1989), and are generally the first bacteria to regrow from biofilms in drinking water in the absence of a disinfectant residual. *Aeromonas hydrophila* is recognized as an organism of concern in drinking water (Reasoner, 1991), and its presence in water has been correlated with childhood diarrhea and the detection of clinical isolates (Moyer, 1987; Burke et al., 1984). The density and types of species appears to depend on the degree of eutrophication in a body of water (Araujo et al., 1991), so the presence of certain aeromonads may be a useful indicator of the microbial quality of drinking and receiving waters.

1.1.2 Detection of Pathogenic Bacteria

Almost 25 years ago Dutka and Bell (1973) pointed out the need for quantitative techniques for the enumeration of small numbers of enteric pathogens in large volumes of wastewater and receiving waters. Due to the relatively low concentrations of most pathogens in surface water, concentration procedures are required. The most common concentration procedures are centrifugation and filtration. Both have advantages and limitations but a filtration approach offers greatest flexibility and simplicity and best meets the criteria of both low initial and operating cost. Simple large diameter cellulose acetate or polycarbonate filters can be used to concentrate water. The filters are then dissolved in acetone or phenol and nucleic acids are extracted from the entrapped organisms.

While this technique is simple and suitable for PCR analyses, it cannot be used for culture-based tests. Levin et al., (1974) described an approach utilizing small Balston filters that were immersed into a water sample, and water was drawn through the filter by vacuum. A series of Balston tubes was used and each tube was placed into an appropriate medium and analyzed by most probable number technique. The authors were able to analyze 10 L in 60 minutes, but the process appeared somewhat cumbersome. Presnell and Andrews (1976) described a membrane filter-most probable number procedure that utilized Filter Aid to increase the amount of water that could be passed through the membrane filter. The filters were washed in a diluent in a blender and the analysis conducted by MPN procedure. They concentrated bacteria from 4 to 5 L of surface water. Van Sluis and Yanko (1978) developed a concentration procedure at the County Sanitation Districts of Los Angeles County (CSDLAC) laboratory to study the occurrence of *Salmonella* in disinfected effluents and receiving water. The procedure utilized

Whatman GF/F glass fiber filters overlaid with diatomaceous earth (DE) to concentrate 20-L samples of surface water using a pressure filtration apparatus. The glass fiber filter and DE was then emulsified in diluent. *Salmonella* spp. were detected at concentrations of 0.06 MPN/liter. These various approaches suggest that reasonably simple procedures can be used to increase the potential detection of pathogenic bacteria in water.

Conventional methods for recovery of pathogens have not changed significantly in the past 30 years. No single procedure is available for reliable detection of any pathogen or group of pathogens. Because the presence of pathogens is intermittent and the survival times in the environment are variable, routine examination of water or wastewater is generally not attempted. Even when disease outbreaks occur, the recovery of pathogens from water is limited by lack of facilities, untrained personnel, inadequate methods, and high costs. There is little to suggest that conventional, culture-based analyses can be used for compliance monitoring or to routinely assess the potential presence of pathogenic bacteria, although conventional analysis for some pathogens is of value in specific situations. Fluorescent antibody techniques have been employed for direct microscopic enumeration of some pathogens in environmental settings (Ginsberg, et al., 1972; Howgrave-Graham and Steyn, 1988), but experience in the CSDLAC laboratory suggested that cross reactivity can be a major problem with wastewater effluent samples (unpublished data).

1.1.3 Microbial Indicators

The many limitations involved with conducting conventional pathogen analyses lead to the use of surrogate organisms to predict the likely presence, or conversely, absence of pathogenic bacteria in water. Because most common waterborne diseases are enteric, coliform bacteria became widely accepted as a suitable indicator of water contamination from soil or fecal pollution and the inherent risk of contamination. Alternative surrogates have been proposed to replace coliforms, and some epidemiological evidence suggests enterococci and *E. coli* can better predict health risk than the broader coliform group (Cabelli, 1982). The health risks observed, however, appear to be viral-associated, and it is unknown if these surrogates provide equivalent prediction of the presence of pathogenic bacteria. *Clostridium perfringens* (Payment and Franco, 1993; Chauret et al., 1995) and male-specific (FRNA) coliphage (Havelaar, 1993) have been proposed as alternative indicators, some data suggest these are better surrogates for protozoan cysts and viruses respectively, but there is little information relating their presence to specific pathogenic bacteria. A study of an estuarine system affected by sewage during rainfall events indicated that *C. perfringens* spores were the best indicator of fecal pollution and were the only indicator that significantly correlated with any of the pathogens in the water column (*Giardia* spp. and *Aeromonas* spp.) (Ferguson et al., 1996). Male-specific coliphages did not correlate with any of the pathogens studied. In contrast, it has been reported that male-specific coliphages were a reliable indicator of fecal contamination in source waters and were also a useful indicator of the ability of drinking water treatment processes to reduce human enteric viruses in water (Sobsey et al., 1995). Despite conflicting reports of their utility, the basic reasons for using surrogates remain, and an indicator that provides a more reliable assessment of the potential presence of bacterial pathogens would be of value to the water industry in general.

1.1.4 Molecular Detection Methods

It is now generally accepted that traditional enrichment and isolation techniques reflect only a small proportion (possibly less than 1%) of the total bacteria actually present in a particular environment. Many environmental bacteria are too starved, stressed, injured, or simply not metabolically adapted to grow on the nutrient rich media typically used for bacterial cultivation. Given that conventional culture-methods for pathogenic bacteria are relatively slow, time consuming, and for some organisms inefficient and technically difficult, molecular approaches which do not rely on culturing bacteria have been employed. Over the last 15 years, the application of molecular methods to environmental samples has transformed our understanding of the composition, phylogeny, physiology, and function of microbial communities in the environment. These methods have led to an increased understanding of microbial processes and revealed the incredible diversity of microbial communities in natural environments. The detection and identification of nucleic acids are useful tools for characterizing microorganisms in natural habitats, and many researchers have taken advantage of the sensitivity, specificity, and rapidity of molecular techniques to detect specific microorganisms in various types of environmental samples.

The earliest application of molecular methods for the detection of specific pathogens in environmental samples involved the use of pathogen-specific probes comprising cloned fragments of DNA complementary to genomic sequences within the target bacteria. However, the methods required lengthy and time-consuming protocols, were difficult to optimize with respect to specificity, and could only detect microorganisms that were present in relatively large numbers. The introduction of methods for in-vitro amplification of specific, defined fragments of nucleic acid circumvented many of the problems of the earlier molecular methods; these methods were quickly adopted by environmental microbiologists.

1.1.4.1 The Polymerase Chain Reaction

The polymerase chain reaction (PCR) is the most commonly used nucleic acid amplification method for generating easily detectable quantities of DNA from very low initial concentrations. PCR is an in-vitro, primer-dependent, enzymatic process that rapidly generates multiple copies of user-specified, defined fragments of DNA (Mullis et al., 1994). With the latest instrumentation, up to a million-fold amplification of target DNA sequences can be obtained in 30 minutes. The resulting amplicons are typically detected by agarose gel electrophoresis and confirmed by hybridization with specific oligonucleotide probes after blotting onto nylon membranes. Positive identification consists of visual detection of an appropriately sized amplicon on the electrophoresis gel and a positive reaction with a probe. This approach is considered a heterogenous method since the amplification and detection assays are performed separately.

The application of PCR to the detection of bacterial pathogens in environmental samples involves concentration of the microbial fraction in the particular sample (usually by centrifugation or filtration), extraction of nucleic acids from the microbial pellet by physical and/or chemical means, and finally, targeted amplification of pathogen-specific nucleic acid fragments by PCR. Providing that the PCR primers are correctly designed and the reaction parameters are tightly controlled, only samples containing the pathogen of interest should result in positive detection. The specificity of the DNA primers used for PCR-based amplification can be varied by careful selection of the target nucleic acids such that the reaction can amplify DNA from all bacteria, a functional group of bacteria, an entire genus, a single species, or individual pathogens. The principles of primer design have been described in detail elsewhere (Giovannoni, 1991; Marchesi, 2001).

One of the most commonly applied molecular techniques in environmental microbiology during recent years has been the phylogenetic analysis of 16S rRNA gene sequences amplified by PCR from DNA (or RNA) extracted directly from a wide range of terrestrial and aquatic environments, using universal bacterial primers targeting most species of bacteria. Since the approach does not rely on culturing microorganisms, it is not subject to the biases and limitations of the agar plate. While many of the bacterial 16S rRNA gene sequences recovered from environmental samples using this approach display a high degree of similarity to cultured and identified bacteria, a large number of sequences are recovered that do not closely match currently identified bacteria, and many "novel" sequences with very little homology to any recognized bacterial species or to the uncultured bacterial sequences recovered in other 16S rRNA-sequencing studies, have also been recovered (DeLong et al., 1993; Field et al., 1997; Fuhrman et al., 1993; Pace, 1997; Rochelle et al., 1994; Rochelle, 2001a). This application of 16S rDNA amplification and sequencing to environmental samples has revealed a vast and previously unrecognized microbial diversity, and indicates that the majority of microbial life on earth has yet to be cultivated. There are many major division and subdivision-level groups of bacteria that exist only as clusters of 16S rRNA sequences. In a more narrowly focused application of the technology, examples of specific functional groups of bacteria targeted by PCR-based amplification include ammonia oxidizing bacteria (Sinigalliano et al., 1995), carbon dioxide fixing bacteria (Xu et al., 1996), mercury resistant bacteria (Bruce et al., 1995), and bacteria carrying dehalogenase enzymes (Marchesi, 2001), all of which may represent cross-species and cross-genus groups of bacteria. The most focused examples of the application of PCR to environmental samples are the detection and identification of specific pathogens such as *Cryptosporidium*, *Giardia*, enteric viruses, and toxigenic bacteria (Table 1-1).

Table 1-1. Detection of Pathogenic Microorganisms in Environmental Samples Using the Polymerase Chain Reaction.

Pathogen	Sample type	Reference
Bacillus anthracis	Soil	Patra et al., 1998
Bacteroides spp.	Water/Wastewater	Brunk and Li, 2001
Caliciviruses	Water	Huang et al., 2000
Campylobacter	Environmental waters	Sails et al., 2002
Cryptosporidium	Water	Rochelle, 2001b
Cyanobacteria (toxic)	Water	Rudi et al., 1998
Cyclospora	Wastewater	Sturbaum et al., 1998
Enteroviruses	Surface waters	Chapron et al., 2001
Escherichia coli (toxigenic)	Natural waters	Lang et al., 1994
Giardia	Water	Abbaszadegan et al., 1997
Helicobacter pylori	Water	Enroth and Engstrand, 1995
Legionella pneumophila	Drinking water	Wellinghausen et al., 2001
Microsporidia	Water	Dowd et al., 1998
Mycobacteria	Water, Sediment, Vegetation	Marchesi, 2001
Reoviruses	Surface water	Spinner and DiGiovanni, 2001
Yersinia enterocolitica	Water	Kapperud et al., 1993

This last application has found widespread popularity because of the many potential benefits of PCR-based detection over more traditional culture- or microscopy-based methods. These benefits include:

1. Tunable specificity: reactions can be controlled such that the specificity ranges from individual pathogens to all bacteria, as described above.
2. Extreme sensitivity: the technique has the potential to detect single cells (although this potential is frequently not achieved with environmental samples).
3. High throughput: the technique is readily amenable to automation, allowing multiple samples to be processed simultaneously.
4. Rapid analysis: the latest instrumentation can complete up to 96 reactions, including probe confirmation, in less than two hours.
5. Multiplexing: Multiple pathogens can be targeted simultaneously in a single reaction.
6. Detailed genetic analysis: the DNA product of the reaction can be sequenced and the resulting sequence used for conclusive confirmation of pathogen identity and molecular "fingerprinting" for epidemiological and source tracking purposes.

Although PCR was seen by many as a panacea for the detection of pathogens and other microorganisms of interest when first developed, it is now known that, in contrast to the advantages described above, the technique can introduce biases and errors into the results of microbial analysis. The application of PCR to environmental samples can be particularly problematic and the various issues concerning PCR inhibition, contamination, primer biases, and other potential limitations have been discussed in detail previously (Hugenholtz and Goebel, 2001).

Despite such limitations and the difficulty in adapting PCR to environmental samples, recent PCR-based studies have dramatically altered the water industry's understanding of many pathogens and led to a much broader level of knowledge concerning pathogen detection and source tracking. Examples of such pathogens include *Cryptosporidium parvum* and *E. coli* O157:H7. Based on sequence analysis of PCR-derived amplification products and other methods of molecular characterization, it is now recognized that *C. parvum* is comprised of at least two distinct genotypes or sub-species with different host ranges (Rochelle, 2001). The anthroponotic genotype (genotype 1) has been isolated almost exclusively from humans while the zoonotic genotype (genotype 2) has been detected in humans and a wide diversity of other animals. A compilation of genotyping data from 22 outbreaks demonstrated that 67% of U.S. outbreaks were caused by the anthroponotic genotype, indicating a human source of contamination (Rochelle et al., 2001c). There is also increasing evidence that sub-genotypes exist within each genotype of *C. parvum* and that some of the sub-genotypes were restricted to specific geographic locations. *C. parvum*-specific PCR-based analysis of stormwater runoff samples indicated human and farm animal waste could be excluded as the source of contamination leaving various kinds of wildlife as the most likely source(s) (Xiao et al., 2000). Consequently, PCR has not only been used for detection of *C. parvum* in water but also to aid in tracing the source of contamination.

A second example of the utility of PCR is provided by enterotoxigenic *E. coli*. Standard methods currently used by the water and wastewater industries for monitoring the microbial quality of water involve the detection and enumeration of total coliforms, fecal coliforms, and *E. coli* using traditional culture-based methods. The presence of *E. coli* is confirmed by culturing in media containing methylumbelliferyl-glucoronide (MUG) and relies on the presence of the enzyme β-glucoronidase in *E. coli,* which converts MUG to a fluorogenic product, 4-methylumbelliferone. However, most isolates of *E. coli* O157:H7 do not possess a functional β-glucoronidase gene (Strockbine et al., 1998) and therefore would not be detected by this method. In simple terms, a water sample containing enterotoxigenic *E. coli* O157:H7 would probably result in an *E. coli*-negative result with MUG-based testing methods. In contrast, PCR-based approaches can be designed to target all enterotoxigenic coliforms (*E. coli* and other coliforms that carry the shiga-like toxin genes) by amplifying a fragment of the toxin-encoding genes or specifically targeting *E. coli* O157 by amplifying fragments from the O-antigen biosynthesis genes.

1.1.4.2 Improved Molecular Methods

The heterogeneous method described earlier, which involves separating amplification and detection assays into separate procedures (PCR, agarose gel electrophoresis, blotting, and confirmatory probing) is simple and relatively inexpensive. However, it is time-consuming, labor-intensive, and prone to carryover contamination because of the multiple manipulations of amplification products. Homogeneous methods allow for the specific detection of amplification products directly within the reaction tube without any physical separation and therefore circumvent many of the problems associated with methods used in a heterogeneous format. Such homogeneous methods that can identify the specific amplicon are preferable due to their speed, simplicity, and greatly reduced risk of carryover contamination.

A variety of homogeneous assay formats have been developed for PCR. The simplest are those that do not include a probe but measure the yield of amplification products by incorporation of a fluorescent dye into the amplicon or by use of a fluorescently labeled primer. While such methods allow homogeneous detection of the amplification product, they cannot confirm the identity of the amplicon. Incorporating a suitably labeled nucleotide probe directly into the assay permits both "real-time" acquisition of PCR signal and simultaneous probe confirmation of amplicon identity. These "real-time" PCR methods use a variety of probe formats, including single fluorescently labeled probes; fluorescent resonance energy transfer probes in which two probes are used, each labeled with a different fluorophore; linear TaqMan probes that carry fluorescent and quencher moieties on either end of one probe; and molecular beacons that also contain fluorescent and quencher molecules on the same probe. The term "real-time" as applied to these technologies means that the assay products (amplified DNA) are detected while the amplification is still in progress. It does not provide real-time detection of pathogen DNA (or any other nucleic acid of interest) in environmental samples. The advantages of real-time methods include:

1. As described above, real-time PCR-based assays are performed in a closed, homogeneous format; consequently, the potential for amplicon contamination is low compared to conventional (heterogeneous format) PCR.
2. Because the amount of amplification product is quantitated as the reaction progresses, the time required for real-time assays is much shorter than conventional PCR, which requires post-reaction analysis that can take up to 24 hours.
3. The quantitation range for real-time PCR assays is much greater (5-6 logs) than for conventional PCR methods that rely on gel electrophoresis and probing (2-3 logs).

Molecular beacons are single-stranded nucleic acid molecules that possess a stem-and-loop structure with the loop containing a probed sequence that is complementary to the target sequence (Tyagi and Kramer, 1996). The hairpin stem is formed by annealing of the two complementary arm sequences that are not related to the target. A fluorogenic molecule is attached to the end of one arm and a nonfluorescent quenching moiety is attached to the other end. In free solution, molecular beacons do not fluoresce because the stem structure keeps the fluorescent molecule close to the quencher and the fluorescence energy is absorbed and released as heat. However, in the presence of target DNA, the loop sequence anneals to the target. forming a probe-target hybrid undergoing a spontaneous conformational change that forces the arm sequences apart, removing the fluorophore from the influence of the quencher and allowing fluorescence emission. Molecular beacons are added to the amplification assay mixture before carrying out amplification, and fluorescence is measured in real-time. Because non-hybridized molecular beacons do not fluoresce, the probe target hybrids from other reaction components prior to signal acquisition. When present in sufficient concentration, molecular beacons interact rapidly with target amplicons, generating fluorescence in the process. The stem region opens at high temperature, and the beacons anneal with the target strands rather than self hybridization. As the target strands synthesized in a reaction accumulate, the fraction of molecular beacons that are bound to targets increases, causing a brighter fluorescent signal. The many advantages of homogeneous format amplification assays such as PCR-molecular beacons has led to their use for the specific detection of a wide range of human pathogens (Table 1-2).

Table 1-2. Pathogens Detected by Homogeneous Format (Real-Time) Amplification Assays.

Pathogen	Probe format[(a)]	Sample type	Reference
Legionella pneumophila	FRET	Potable water	Wellinghausen et al., 2001
Campylobacter jejuni	TaqMan	Human feces	Cooper et al., 2002
Vibrio parahemolyticus	FRET	Food	Iwade et al., 2002
Listeria monocytogenes	TaqMan	Water	Nogva et al., 2000
Mycobacterium avium	Molecular beacon	Bovine feces	Fang et al., 2002
West Nile Virus	Molecular beacon[(b)]	Mosquitoes	Lanciott and Kerst, 2001
Encephalitozoon intestinalis	FRET	Human feces	Wolk et al., 2002
Cryptosporidium parvum	TaqMan	Source waters	Abbaszadegan et al., 1998

(a) Assays based on PCR amplification combined with the various probe formats: FRET, fluorescence resonance energy transfer; TaqMan, dual labeled probes used in 5'-nuclease assay.
(b) Nucleic acid sequence-based amplification (NASBA) used instead of PCR.

1.2. Research Scope

Current methods and standards for the detection of coliform bacteria in receiving water may be inadequate predictors of the presence of bacterial and other enteric pathogens. Detection of frank bacterial pathogens is not done routinely, due in part to the lack of accessible technology. Many significant bacterial pathogens are regarded as fastidious organisms, requiring long and specialized incubation periods and media. Microbiological methods currently available do not address the common needs of the wastewater and drinking water industries in their ability to detect either indicator organisms or frank pathogens. A rapid and simple method is needed for the collection, concentration, extraction, and identification of pathogenic bacteria in wastewater effluents, surface water, and finished water. The method should be simple and cost effective to allow water utilities and commercial laboratories to perform the entire test, or partially process the sample and send it in a stabilized form to a reference or commercial laboratory.

A simple concentration method combined with total DNA extraction and specific and sensitive PCR amplification potentially represents a rapid and cost-effective means to monitor receiving waters for multiple bacterial pathogens. Sample volumes ranging from 100 mL to 10 L can be concentrated by filtration onto 47-142 mm diameter membranes or glass fiber filters. The retained bacteria can be eluted and assayed by conventional culture methods or their DNA and RNA directly extracted and purified for PCR amplification and probe hybridization. The advantage of this approach is that it can potentially be applied for the rapid detection of *E. coli, Salmonella, Shigella, Campylobacter jejuni, Helicobacter pylori, Aeromonas* species, *Yersinia, Mycobacterium avium* complex (MAC) organisms, and other bacterial pathogens retained in the membrane or filter. Amplification primers and specific oligoprobes are available for the detection of these organisms. Several methods for isolation of total DNA from sediments, soil, foods, and other materials have been developed (Tsai and Rochelle, 2001) and adapted for use with surface water samples. In addition, commercial kits for total DNA extraction and purification resulting in PCR-compatible samples arc available.

This project attempted to address critical needs in the detection of bacterial pathogens from receiving waters including: 1) simple concentration methods for bacteria in 100 mL to 10-L volumes, compatible with conventional culture methods and molecular techniques; 2) specific and rapid detection of fastidious, noncultural and slow growing bacteria in a short period of time; 3) simultaneous sampling and detection of multiple bacterial pathogens; 4) a novel approach for rapid and specific probe hybridization of amplified samples; and 5) an evaluation of the utility of alternative indicators to predict the probability of pathogens in receiving water. Results of this project may benefit not only the wastewater industry's testing laboratories but also help improve the operation of wastewater treatment plants, which can be translated to more effective promotion of public health. This project may provide the water industry with tools and information to more accurately assess the risk to public health posed by waterborne pathogens and aid the industry in planning watershed management strategies.

1.3 Project Objectives

The overall objective of this project was to examine the feasibility and the validity of an alternative assay to coliform testing for assessing the potential presence of bacterial pathogens in receiving waters. PCR detection of pathogenic bacteria was evaluated both as a research tool to assess the predictive value of proposed alternative indicators and as a primary simplified analytical method for bacterial pathogen detection. A potential PCR method using molecular beacon technology was also evaluated and optimized for the rapid amplification and detection of two bacterial pathogens of significant public health concern. Specific research tasks to address this objective included: 1) testing and optimizing simplified sample concentration and elution/extraction methods for bacterial pathogens in 100 mL to 10 L of water; 2) comparing a combination of PCR technology and conventional culture methods to detect toxigenic *E. coli*, *Salmonella*, *Shigella*, *Campylobacter*, *Helicobacter*, *Mycobacterium*, *Yersinia*, and *Aeromonas* and the surrogates *E. coli*, enterococci, *Clostridium perfringens,* and male-specific coliphages from water to determine the relative predictive value of the surrogates; 3) conducting a large-scale field survey to determine the prevalence of indicators and pathogens in wastewater effluents and in a wide diversity of surface waters; and 4) assessing the potential to transfer the selected sample concentration method and surrogate, or molecular-based technology, from the research environment to routine analytical settings in water laboratories with limited molecular experience.

CHAPTER 2.0

PHASE 1 DEVELOPMENT AND EVALUATION OF METHODS

2.1 Evaluation of GF/F-DE Concentration Method

The proposed method for concentrating bacteria from surface water samples (and treated effluent) was to filter the water sample, using a pressure filtration apparatus, through a 47 mm Whatman GF/F glass fiber filter overlaid with 1.4 g diatomaceous earth (DE). Advantages of this approach are 1) the filter fibers and DE can be dissociated in a diluent, releasing the entrapped bacteria, 2) the filters and materials are inexpensive, 3) the necessary equipment is readily available and simple to operate, and 4) the procedure is potentially adaptable to field work. The CSDLAC laboratory had previously used this procedure to successfully concentrate up to 30 L of water to measure salmonellae concentrations in disinfected effluent and surface water in a Southern California creek (Van Sluis and Yanko, 1977). Experiments with coliform bacteria demonstrated that 99.9% of the bacteria were retained by this filtration system. While the method was simple and effective at concentrating bacteria, it had not been evaluated either with bacteria other than coliforms and *Salmonella* spp. or for use in conjunction with molecular detection methods.

The first step of this project was to document that the bacterial capture rate observed in the 1977 work could be duplicated and to evaluate the concentration method with a larger range of water qualities and other target organisms.

2.1.1 Description of 47-mm and 142-mm GF/F-DI Filtration System

Two filtration systems were evaluated. The first, used for the preliminary experiments, consisted of a Pall Gellman 47-mm stainless steel pressure funnel (VWR Cat. #28144-652) used in conjunction with a Whatman GF/F 47-mm glass fiber filter (VWR Cat. #28497-958) overlaid with 1.4 g of diatomaceous earth (DE) (Spectrum 68855-54-9). The 1.4 g DE was suspended in 100 mL deionized water (DI) and sterilized by autoclaving. The sterile DE suspension was poured into the filter holder reservoir and filtered through the GF/F filter before the sample was pumped through the filter holder. The sample was pumped from a 10-L carboy through the pressure filtration vessel using a large diaphragm pump (Milton Roy, Model No. FR141-144) that was available at the beginning of the study.

The second filtration system was conceptually similar but used a larger 142-mm large-capacity 316 stainless steel pressure holder with a 1.5 L-reservoir and pressure relief valve set at 100 psi (Whatman Cat. #1825142). The filter holder was loaded with a 142-mm GF/F filter (VWR Cat. #2847-969) and overlaid with 10.8 g of DE (Spectrum 68855-54-9). The 10.8g DE was added to approximately 1 L of DI water and autoclaved to sterilize. The sterile DE was resuspended in the liter of water and poured aseptically into the filtration vessel through a screw cap inlet on top of the reservoir. That liter of water with the DE acted as a "cushion" when sample filtration was started to prevent disruption of the DE layer on the surface of the GF/F filter. The 1 L approximate volume of liquid was maintained over the filter/DE through the entire sample filtration. After the initial experiments, a smaller light-weight diaphragm pump (Primus 221, purchased from ALLDOS, No. 221-27) replaced the Milton Roy pump and was used to pump the sample through the 142-mm pressure filtration apparatus for the remainder of the study. Both the 47-mm and 142-mm filtration systems are shown in Figure 2.1.

Filtration was terminated when the 10-L sample reservoir ran dry. If the filter clogged, filtration was stopped, and the remaining sample was aseptically removed from the filtration vessel. Another filter and fresh DE were added, and the filtration was continued until complete. Only the dirtiest, most turbid samples required a second filter to complete filtration of 10-L sample volumes.

After the sample was filtered, the pressure filter was opened, and the GF/F filter with DE was aseptically removed and placed into a sterile wide-mouth plastic jar (Nalgene 500 mL, VWR Cat. No. 36319-707), as illustrated in Figure 2.1. The appropriate selective or nonselective enrichment broth was added to the wide mouth jar, shaken, and incubated per the specified protocol as described in Appendix A and Section 3.3.

Between samples the 47-mm pressure filtration apparatus was sterilized by autoclaving. The larger 142 mm pressure filtration vessel was disassembled, exposed surfaces covered with aluminum foil, and autoclaved. The pump and associated tubing was disinfected by recirculating a solution of chlorine and Liqinox. The disinfecting solution was made by adding 200 mL of chlorine (Chemchlore, Home Depot) and 10mL of 5% Liquinox to approximately 3 L of tap water. The pump was started and, when time permitted, the chlorine solution was recirculated overnight. Minimum disinfection time between samples was two hours. The pump was dechlorinated by first dissolving 70 g of sodium thiosulfate in 2-2.5 L of DI water and recirculating for five to 10 minutes, after which the pump and tubing were rinsed with 10 L of sterile DI water.

2.1.2 Retention of Bacteria by Glass Fiber Filter with Diatomaceous Earth

The initial bacterial retention experiments were conducted by filtering 10-L water samples through the 47-mm filtration apparatus. Bacterial retention was determined by measuring the concentration of heterotrophic plate count bacteria (HPC) in the water sample before filtration and in the filtrate. Samples tested included nondisinfected secondary wastewater effluent, tertiary wastewater effluent, and surface water samples. Tertiary effluent samples were dechlorinated and spiked with a small amount of secondary effluent to increase the HPC.

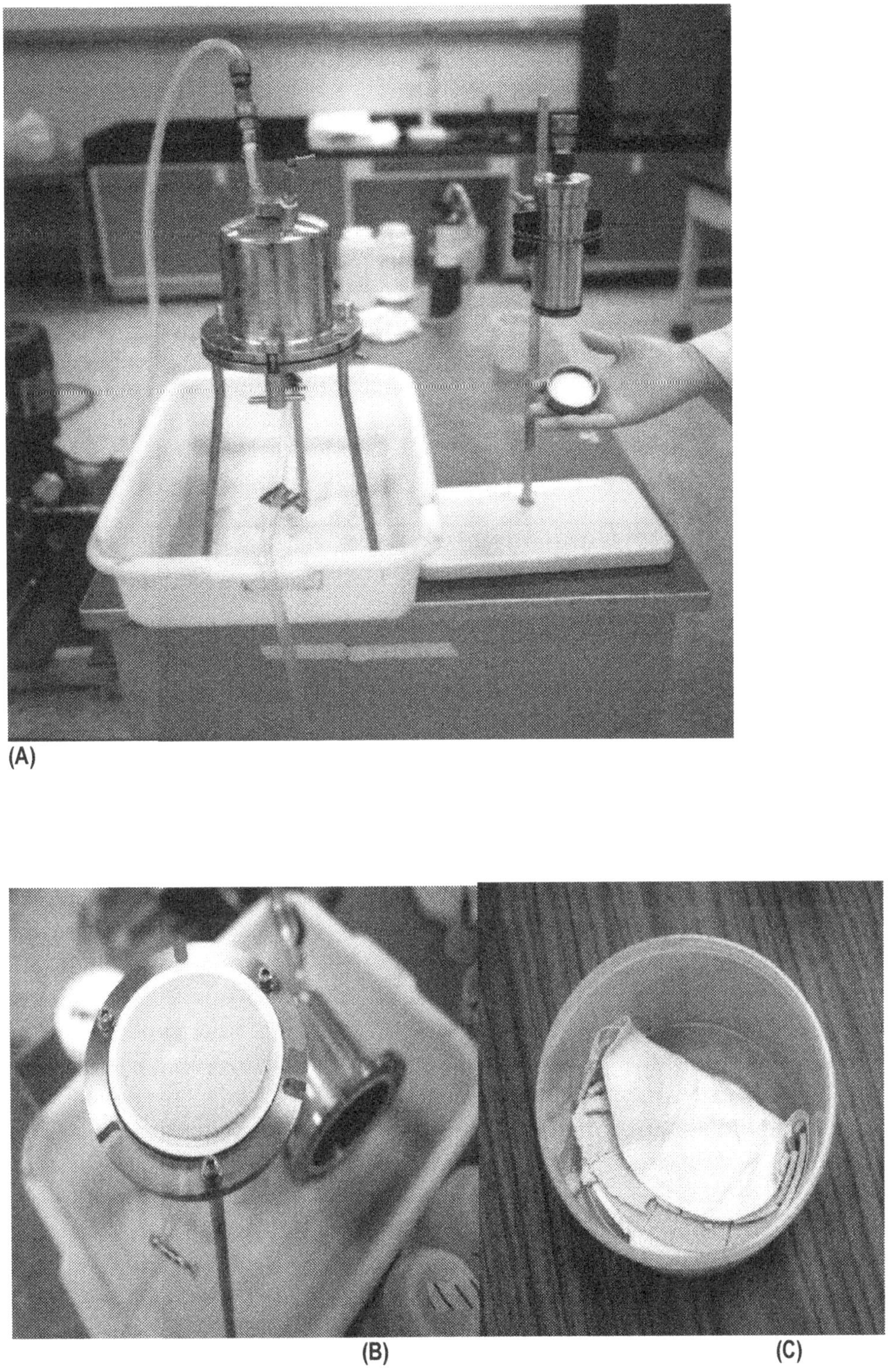

(A)

(B) (C)

Figure 2-1. GF/F-DE Filtration System.
(A) 142-mm and 47-mm pressure filters. Bottom plate removed from 47-mm filter holder showing filter in place. (B) 142-mm holder opened after sample filtration showing GF/F filter with DE layer on top. (C) GF/F filter with DE placed in plastic wide-mouth jar. Selective or nonselective enrichment broth is added into jar, contents shaken, and then incubated.

Turbidity of the water samples tested ranged from <1 to 23. During the initial retention experiments, it was found that the 10-L sample volume goal for this study could not always be achieved using the 47-mm filtration apparatus when sample turbidities were high. Experiments were conducted to determine the relationship between turbidity and the filter surface area required to reliably concentrate 10-L sample volumes (described in Section, 2.1.3). Based on these experiments, it was determined that a 142-mm GF/F filter should be able to handle 10-L sample volumes for the range of turbidities anticipated in surface water samples. Bacterial retention experiments were then repeated using the 142-mm filtration apparatus to confirm that retention was equivalent to the smaller filtration system. Bacterial retention for both the 47-mm and 142-mm filtration systems is shown in Table 2-1. Retention on the 47-mm filters was equivalent to that observed in 1977 study (VanSluis and Yanko), and retention on the 142-mm filter was slightly lower but adequate to continue with sensitivity experiments to document the utility of the method for recovering specific target organisms.

Table 2-1. Retention of Heterotrophic Plate Count Bacteria in 10-L Wastewater Effluent and Surface Water Samples on 47- and 142-mm GF/F Filters with DE.

Water Sample	Turbidity - NTU Mean & (Range)	47 mm GF/F-DE n[(a)]	47 mm GF/F-DE % Retention	142 mm GF/F-DE n[(a)]	142 mm GF/F-DE % Retention
Wastewater Effluents	9 (1.3 - 4.5)	7	99.99	-	-
Surface Water	3 (<1 - 23)	1	98[(b)]	3	99.8

(a) Number of analyses.
(b) Represents one sample with turbidity of 23 NTU (nephelometric turbidity units).

2.1.3 Relationship Between Sample Turbidity and Volume Filtered

The Whatman GF/F 47-mm glass fiber filter with a thin layer of DE over its surface permitted the rapid filtration of reasonably large volumes of water using a very simple filtration system. However, during the preliminary experiments to confirm bacterial retention, a limit was found to how much water could be filtered with the 47-mm filter, depending on the turbidity of the sample water. The goal of the project was to be able to concentrate 10-L volumes of water with a detection sensitivity of <10 organisms in a 10-L sample. Experiments were conducted to determine the relationship between turbidity and filter surface area needed to assure the ability to concentrate a 10-L sample. These experiments were conducted using the 47-mm filtration apparatus, and results were extrapolated to estimate the filter size needed for water qualities ranging from pristine to stormwater runoff.

Two water samples were used to develop data on the relationship between sample turbidity and the volume of water that could be filtered through the 47-mm GF/F-DE filter. The first sample was river water from a local river in the Los Angeles area; the sample collected had a turbidity of 4.1 NTU. The second sample was primary effluent from a wastewater treatment plant, which had a turbidity of 29.3 NTU. A series of dilutions was made from each sample, using laboratory ultrapure water, to generate a series of 10-L volume samples with varying turbidity. Each of the resulting samples was filtered through the 47-mm filtration apparatus until the pressure reached 100 PSI and flow through the filter was stopped. No further dilutions of sample were filtered once an entire 10-L sample volume could be filtered. The data from both sample types were combined and are presented graphically in Figure 2-2.

Based on these data, it was determined that the upper limit for the 47-mm GF/F-DE filter system was a turbidity in the range of 0.6 – 1.3 NTU to reliably concentrate a 10-L volume. A 142-mm GF/F-DE filter was estimated to be able to concentrate 10-L sample volumes with turbidities ranging up to 16 –17 NTU. That was considered acceptable for the majority of anticipated water samples, and based on that estimate, the 142-mm filter was adopted for the remainder of the study. Nevertheless, the 47-mm GF/F-DE filtration system was entirely suitable for concentrating 10-L samples of low turbidity waters, and for exceptionally clean water the 47-mm GF/F-DE filter could be adapted to conventional vacuum filtration filter holders, commonly used for membrane filtration tests, to concentrate multi-liter sample volumes.

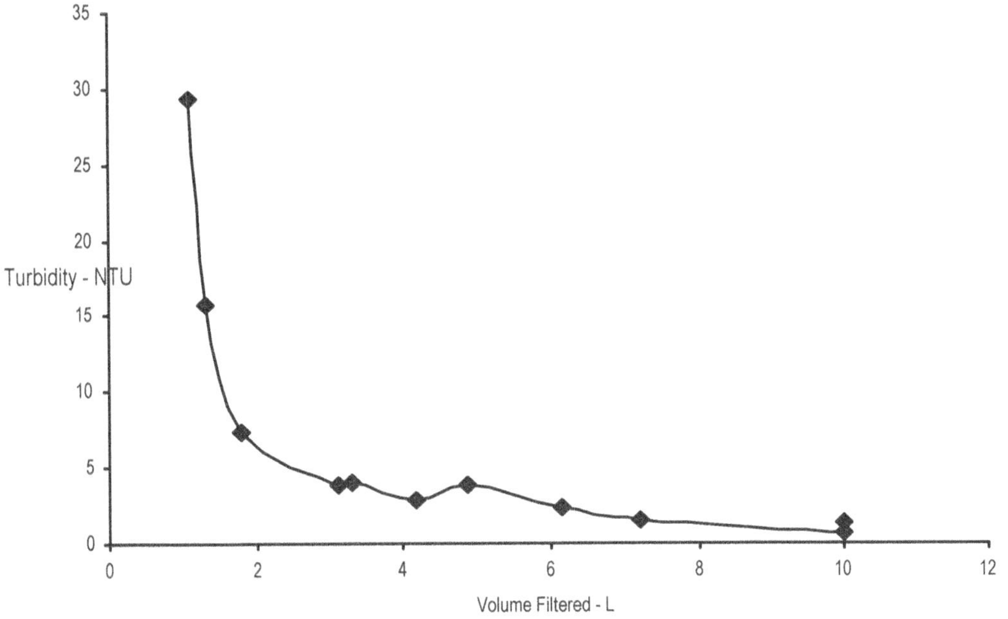

Figure 2-2 Relationship Between Sample Turbidity and Volume Filtered with 47- mm GF/F-DE Filter.

2.1.4 Target Organism Sensitivity Using GF/F-DE Filtration System

The primary goal of the large-volume GF/F-DE concentration method was to concentrate 10-L sample volumes with a detection sensitivity of <10 organisms per 10 L. A series of spiking experiments was conducted to evaluate the sensitivity of the method. The previous experiments determined that bacteria were adequately retained by the GF/F glass fiber filter-DE matrix. Sensitivity, however, is directly related to the ability to recover the target organism from the resulting concentrate. Recovery is dependent on two factors. First is the effect the concentration system itself has on the target organism. It may not be of value to capture the organism if the capture process causes the target organisms to die. Second, assuming the captured organisms remain viable, a culture-based analytical test must be capable of detecting the target organism from a complex mixture of organisms. The target organism may not have to remain viable to be detected using molecular detection methods, but the question of viability is clearly important from other perspectives and is considered throughout this report.

The experimental approach used to validate the sensitivity of the GF/F-DE concentration method was to seed incremental doses of selected target organisms into 10L water samples collected from different sources and determine the detection sensitivity. The seeding doses per 10-L sample were initially 1,000 CFU, 100 CFU, 10 CFU, and 1 CFU. The 1,000 CFU dose was discontinued after the initial series of experiments.

The first seeding experiment was carried out by seeding 10 L of dechlorinated tap water with serial dilutions of *Enterococcus,* ranging from 1,000 CFU to 1 CFU. It was necessary to document the relative error associated with the stated dose, especially at the low doses. The ability to reliably seed a 10-L sample at the higher seed levels of 100 CFU and 1,000 CFU presented no problem. However, the low seed levels of 10 CFU and, in particular, 1 CFU were subject to greater relative variability and more likely to be affected by uneven distribution or pipette error. In the case of a 1 CFU dose, there was some probability that any particular aliquot taken from a suspension of such low bacterial density might not contain a bacterial cell. To address this issue, a "high seed" dilution for the 1,000 and 100 CFU doses and a "low seed" stock for the 10 and 1 CFU doses was prepared so the seed dose could be enumerated directly from the stock cell suspension without further dilution. Greater numbers of replicate plate counts of the seed suspensions were prepared for the lower dose levels to get a better measure of relative variability. To measure the 1 CFU dose, 0.1 mL of the "low seed" suspension was pipeted with a micropipeter into each of 20 Petri dishes; for the 10 CFU dose, 1.0 mL of the "low seed" suspension was planted into each of 12 Petri dishes. The 100 CFU and 1,000 CFU doses were determined by plating the "high seed" suspension in triplicate and inoculating 1 mL into the 10-L sample volume. Results for these plate counts are summarized in Table 2-2 and demonstrate the expected dose variability.

The seed dose at the higher levels of 100 CFU and 1,000 CFU varied by 10% or less. The 10 CFU dose ranged from 8 CFU to 15 CFU. As expected, the 1 CFU dose was relatively inconsistent. Out of 19 plates inoculated at this dose, seven were <1 CFU, 11 plates had 1 to 2 CFU and one plate had 4 CFU. This translates to approximately a 60% probability that the 1 CFU dose would contain a bacterial cell. The Table 2-2 set of data indicating the variability of the seed dose is shown for illustrative purposes. A similar set of data was developed for each seeding experiment summarized in Table 2-3 to document that the seed dose was in the desired range. Most important is the data for the 1 CFU dose. In the Table 2-2 example, there was a 60% probability that the 1 CFU dose actually contained a bacterial cell, or conversely, a 40% probability that the 1 CFU dose did not contain a cell. While there was some variation in this probability throughout the series of seeding experiments, the average of all experiments was approximately a 60% chance that the 1 CFU dose contained a target cell. As is noted later, this figure is significant to assessing the overall success of the sensitivity experiments at the 1 CFU dose level.

Table 2-2. Reliability of Seed Dose – Replicate Plate Counts of *Enterococcus* Dilutions Used to Dose 10-L Tap Water Sample.

1 CFU n=20[(a)]	10 CFU n=10	100 CFU n=3	1000 CFU n=3[(b)]
0	8	119	1130
0	10	118	1180
0	10	130	1190
0	10		
0	11		
0	12		
0	12		
1	12		
1	12		
1	15		
1			
1			
1			
2			
2			
2			
2			
2			
4			

(a) One plate lost due to lab accident.
(b) 1 mL dose based on counts from a 10^{-1} dilution.

Each of the 10- samples of water was dosed with enterococci at the specified concentration and pumped through a 47-mm GF/F filter with filter aid. The filter with diatomaceous earth was removed from the filter holder, placed aseptically into Azide dextrose broth, and incubated for 24 hours. After 24 hours, each of the Azide dextrose broth cultures corresponding to 1,000 CFU, 100 CFU, 10 CFU and 1 CFU doses was streaked onto PSE agar to confirm the presence or absence of *Enterococcus*. All subsequent seeding experiments to document the sensitivity of the concentration method were conducted in a similar manner.

In addition to sensitivity, quantitative recovery was also estimated during most of the concentration method validation experiments (Table 2-3). To measure percent recovery, 10 L of sample water was seeded with 10,000 CFU of the target organism, and filtered though either the 47-mm or 142-mm filtration apparatus using the GF/F-DE system. After filtration, the GF/F-DE filter was aseptically removed and placed in a sterile container containing phosphate buffer (Standard Methods) with 0.1% Tween 80. Samples were homogenized using a Polytron tissue homogenizer and analyzed in triplicate using the five-tube, three-dilution most probable number (MPN) method.

Different target organisms and waters were evaluated through the series of experiments, and both the 47-mm and 142-mm filtration systems were tested. The culture-based media systems used to detect the other target organisms listed in Table 2-3 were more complex than that described above for enterococci. When surface waters or effluents were used in the sensitivity and recovery experiments, identification of the target organism was required, in addition to enrichment and primary isolation, to insure the target organism was recovered. The microbiological methods used to recover the target pathogens during the sensitivity and recovery experiments are detailed in Appendix A. Results of the sensitivity experiments and percent recovery are summarized in Table 2-3.

Table 2-3. Summary of Recovery Experiments Validating GF/F-DE Filter System Detection Sensitivity in 10-L Samples.

Organism	Water Source	Filter Size mm	n[(a)]	No. Positive at CFU Dose per 10-L: 1000	100	10	1	Average Recovery: n[(b)]	%
Enterococci	Tap	47	2	2	2	2	2	1	110
Salmonella	Tap	47	1	1	1	1	1	1	16
Salmonella	Effluent	47	1	(c)	1	1	0	1	63
Salmonella	Effluent	142	5	(c)	5	5	1	5	62
Salmonella	USW[(d)]	142	5	(c)	5	4	3	5	32
Salmonella	UIS[(e)]	142	4	(c)	4	2	1	4	43
E. coli 0157:H7	Tap	142	2	(c)	2	2	0	1	54
E. coli 0157:H7	Effluent	142	3	(c)	3	3	1	3	72
E. coli 0157:H7	UIS	142	2	(c)	2	2	1	2	43
Yersinia	Tap	142	4	(c)	4	3	0	4	51
Yersinia	Effluent	47	1	(c)	0	0	0	(f)	(f)
Yersinia	USW	142	3	(c)	0	0	0	(f)	(f)
Yersinia	UIS	142	2	(c)	0	0	0	(f)	(f)

(a) Number of experiments conducted.
(b) Percent recovery was not determined with every sensitivity experiment, so "n" is less in some rows.
(c) The 1,000 CFU per 10-L dose was discontinued after initial experiments.
(d) "USW" indicates untreated surface water of generally good water quality.
(e) "UIS" indicates urban influenced streams of generally poor water quality.
(f) Percent recovery was not considered since no *Yersinia* were recovered in sensitivity experiments.

2.1.5 Discussion of GF/F-DE Filtration System Validation Experiments

The goal of the GF/F-DE validation experiments was to demonstrate that a simple, cost-effective filtration system was able to reliably recover <10 bacteria in a 10-L water sample. The system had to be able to concentrate 10-L samples of water and/or effluent representing a fairly wide range in water quality.

The bacterial retention experiments demonstrated that 98% to 99.99% of the bacteria in a water sample would be retained by this system. However, there are two parts to a concentration system that are suitable for recovering microorganisms: 1) the ability of the concentration system to retain or capture the target organism, and 2) the ability to recover the target organism from the resulting concentrate. A total of 35 sensitivity and recovery experiments are summarized in Table 2-3, confirming that viable target organisms could be recovered from the GF/F-DE concentrates. Actually more than 35 experiments were run. When surface water and effluent samples were seeded and concentrated, a blank (nonseeded) 10-L sample was also concentrated and tested with each experiment. If the target organism was detected in the blank, the experiment was not included in the sensitivity data since it could not be determined if the organism recovered by presence/absence was the added seed or was indigenous in the sample before seeding. Three experiments were not included in Table 2-3 because the target organism was detected in the blank sample. While this reduced the total number of seeded sensitivity experiments reported, it also provided additional evidence that the concentration system was recovering indigenous pathogens from 10-L surface water samples.

The GF/F-DE concentration system proved very sensitive, with the exception of *Yersinia*, which is discussed separately. As defined in this study, sensitivity reflects the ability to recover low numbers of a target organism from a 10-L water sample using a presence/absence assay. Quantitative recovery is the percent recovery of a larger number of the target organism in the original sample. Enterococci, *Salmonella,* and *E. coli* 0157:H7 were recovered in every test at the 100 CFU dose, in 22 of 25 tests at the 10 CFU dose, and in 10 of 25 tests at the 1 CFU dose. At first glance, recovery of the 1 CFU dose may appear relatively poor. However, there was only a 60% average theoretical probability that the samples seeded at a 1 CFU dose actually received a target cell, or only 15 of the 25 samples potentially contained the target cell. Therefore, recovery of the 1 CFU dose was most likely closer to 10 positive out of 15 potentially positive samples. These experiments demonstrated that the GF/F-DE filtration system met the proposed detection sensitivity goals.

Quantitative recovery did not appear to reflect the high degree of sensitivity demonstrated with the low dose experiments. Average recovery (excluding *Yersinia*) of 10,000 CFU of target organism seeded into 10-L samples was 55% and ranged from 16% to 110%, although true recovery is most likely higher. The initial retention experiments showed that in most cases <99% of the bacteria were retained, and the sensitivity experiments showed that very low numbers could be recovered. The quantitative percent recovery is most likely biased to the low side, predominantly due to clumping of cells during the concentration process and aggregation of cells on filter particles and DE. Also, recovery was determined using MPN methods, which further contribute to quantitative variability. Given that the goal was to detect low numbers of pathogens in large volumes of water, the sensitivity data was given greater weight in judging the success of the concentration method.

As seen in Table 2-3, *Yersinia* was not recovered well during the concentration validation experiments. This inability appeared to be due to the failure of available culture-based isolation methods to recover *Yersinia* from the microbially complex matrices present in the 10-L concentrates. The tap water experiments, where background bacterial concentrations were minimal, demonstrated that the GF/F-DE system was able to recover *Yersinia* essentially as well as the other organisms evaluated. When effluent and surface water systems were evaluated, *Yersinia* could not be recovered from the large population of competing bacteria present in the concentrate. Since the goal of the project was to develop and validate a simple, practical, low-cost concentration system and demonstrate its utility for monitoring surface water samples and not to develop new culture-based media systems for pathogens, *Y. enterocolitica* was deleted from the list of pathogens to be analyzed by culture-based methods during the second phase.

In summary, the GF/F-DE concentration system proved to be a simple, low-maintenance, relatively low-cost, and effective concentration system for recovering low numbers of bacterial pathogens from effluent and surface water samples. While some capital investment is necessary to purchase the pressure filtration apparatus, this cost is significantly less than many alternative concentration systems. In addition, these are simple devices and can be expected to have a long service life with minimal care. The disposable materials, the GF/F filters, and DE represent minimal cost. The primary drawback is the complexity and in some cases (*Yersinia*) the failure of available culture methods to recover pathogens from the resulting concentrate. That problem may be alleviated by combining the concentration system with molecular detection methods, as described in following sections.

2.2 Selection and Evaluation of PCR Primers and PCR Conditions

The goal of this project was to apply previously published primers and PCR conditions to environmental samples without extensive optimization or method development. Nonetheless, more work had to be done in this area than originally anticipated because some published primer sets did not perform adequately, were not directed at the most appropriate target sequences for a waterborne pathogen, or did not meet internal quality control and assurance criteria. In order to meet the goals of this project all primer pairs and probes, whether previously published or newly selected within this project, were subjected to the following tests: 1) specificity and compatibility testing by comparison of primer/probe sequences with online DNA sequence databases; 2) empirical verification of efficient amplification with the published PCR conditions with optimization of conditions conducted where appropriate; 3) specificity of amplification for the intended target; and 4) cross-reactivity testing with 44 other organisms representative of environmental water samples. A total of 20 primer sets were evaluated for use in this project (Table 2-4).

Table 2-4. Specific Primers Used for the Detection of Target Organisms.

Organism	Primers	Primer Sequences (5'-3')	Target Gene	Reference
Escherichia coli 0157:H7	SLTII-1 SLTII-2	CTTCGGTATCCTATTCCCGG GGATGCATCTCTGGTCATTG	shiga-like toxin gene	Olsvik et al., 1993
Escherichia coli 0157:H7	O157PF8 O157PR8	CGTGATGATGTTGAGTTG AGATTGGTTGGCATTACTG	rfb gene	Maurer et al., 1999
Salmonella typhimurium	SHIMA-L SHIMA-R	CGTGCTCTGGAAAACGGTGAG CGTGCTGTAATAGGAATATCTTCA	himA	Bej et al., 1994
Yersinia enterocolitica	ail664 ail833	ACTCGATGATAACTGGGGAG CCCCCAGTAATCCATAAAGC	*ail* gene	Bhaduri, et al., 1997
Aeromonas hydrophila	AH16S450 AH16S1135	GAAAGGTTGATGCCTAATACGTA CGTGCTGGCAACAAAGGACAG	16S rRNA	Dorsch et al., 1994
Campylobacter jejuni	PG50 PG03	ATGGGATTTCGTATTAAC GAACTTGAACCGATTTG	flaA	Oyofo and Rollins, 1993
Helicobacter pylori	93275 93275	AAGCTTTTAGGGGTGTTAGGGGTTT AAGCTTACTTTCTAACACTAACGC	URE-C	Lage et al., 1996
Mycobacterium avium Complex	Primer 3 Primer 4	GGGTGACGCGACATGGCCCA CGGGTTTCGTCGTACTCCTT	dnaJ	Takewaki, et al., 1993
Shigella sonnei	SH1 SH2	CTGGATGGTATGGTGAGG GGAGGCCAATTATTTCC	ial locus	Schoolnik, 1993
Clostridium perfringens	PL7 PL3	ATAGATACTCCATATCATCCTGCT AAGTTACCTTTGCTGCATAATCCC	phospholipase C gene	Fach and Popoff, 1997
MS2 Phage	Upstream downstream	CAAGTTGCAGGATGCAGCGCC GGCTCCTTACAGGCAGCCCG	MS2CG ssRNA	Rose, et al., 1997
Universal Bacterial	66F 342R	AACACATGCAAGTCGAA CCACTGCTGCCTCCCGTAG	16S rRNA	Rochelle, et al., 1994

Published PCR conditions were used as a starting point for each of the primer sets selected and optimized as appropriate. Primer annealing temperature and $MgCl_2$ concentration were the primary conditions tested for optimal amplicon production and absence or significant reduction in nonspecific products. Optimized conditions differed from published conditions for *Escherichia coli* 0157:H7, *Salmonella typhimurium*, *Campylobacter jejuni*, *Shigella sonnei*, and MS2 phage (Table 2-5).

Table 2-5. PCR Conditions for Selected Primers.

Organism	Primer Set	Published Conditions	Optimized Conditions	PCR Product
E. coli 0157:H7	SLTII	60°C, 1.5 mM $MgCl_2$	60°C, 3.0 mM $MgCl_2$	482 bp
S. typhimurium	SHIMA	65°C, 2.5 mM $MgCl_2$	60°C, 1.5 mM $MgCl_2$	122 bp
Y. enterocolitica	ail	Not given in paper	58°C, 1.5 mM $MgCl_2$	172 bp
A. hydrophila	AH16S	64°C/60°C[a], 1.5 mM $MgCl_2$		685 bp
Campylobacter jejuni	PG	37°C, 1.5 mM $MgCl_2$	49°C, 2.5 mM $MgCl_2$	450 bp
Helicobacter pylori	Ure-C	55°C, 1.5 mM $MgCl_2$	50°C, 2.5 mM $MgCl_2$	294 bp
M. avium Complex	dnaJ	65°C, 1.5 mM $MgCl_2$		365 bp
Shigella sonnei	SH	43°C, 1.5 mM $MgCl_2$	50°C, 2.5 mM $MgCl_2$	320 bp
Cl. perfringens	PL	55°C, 1.5 mM $MgCl_2$		283 bp
MS2 Phage	MS2ssRNA	50°C, 1.5 mM $MgCl_2$	55°C, 1.5 mM $MgCl_2$	218 bp
Universal Eubacterial	16S rRNA	50°C, 2.5 mM $MgCl_2$		320 bp

(a) 64°C for 1 cycle then 60°C for 39 cycles.

In order to evaluate the specificity of each synthesized primer set, the nucleic acid was extracted from the target organism, amplified, and verified by gel electrophoresis. Amplicon targets were also subsequently verified by probe hybridization (see section on probes below). For bacterial targets, DNA was extracted using a modified Chelex-100 boiling lysis extraction. For RNA bacteriophage MS2, virus particles were partially purified by Sephadex G50 column purification, and RNA was extracted by boiling lysis and immediately converted to cDNA by a reverse transcriptase reaction. Specificity testing indicated that the selected primers effectively amplified their respective targets with the optimized conditions (Table 2-6).

A matrix cross reactivity test was performed with the 12 primer sets against all target organisms and a range of 44 organisms representative of environmental water samples. These organisms were, for the most part, obtained from ATCC, reconstituted, and the DNA extracted from actively growing cultures using the Chelex-100 boiling lysis protocol. Following extraction, the UNIV primer set was used to determine if there was amplifiable DNA from the sample. The details of the cross reactivity testing are tabulated in Appendix B. A positive result (+) indicates that there was sufficient DNA quantity, purity, and integrity present for subsequent PCR amplification using the species-specific primers. The results indicate that under the optimized conditions selected for each primer set, cross reactivity was not observed for any of the primer sets with the nontarget organisms tested.

Table 2-6. Target Organism Specificity.

Organism	Source	Primer Set										
		16S	SHIMA	SLTII	AIL	PL	MS2	SH	UREC	DNAJ	PG	AH16S
Clostridium perfringens	ATCC #13124	+	–	–	–	+	–	–	–	–	–	–
Phage MS-2	ATCC #15597-B1		–	–	–	–	+	–	–	–	–	–
Escherichia coli O157	Clinical	+	–	+	–	–	–	–	–	–	–	–
Salmonella typhimurium [(a)]	ATCC #14028	+	+	–	–	–	–	–	–	–	–	–
Yersinia enterocolitica	ATCC #23715	+	–	–	+	–	–	–	–	–	–	–
Aeromonas hydrophila	Water	+	–	–	–	–	–	–	–	–	–	+
Campylobacter jejuni	ATCC #29428	+	–	–	–	–	–	–	–	–	+	–
Helicobacter pylori	ATCC #43504	+	–	–	–	–	–	–	+	–	–	–
Mycobacterium avium	ATCC #25291	+	–	–	–	–	–	–	–	+	–	–
Mycobacterium intracellulare	ATCC #13950	+	–	–	–	–	–	–	–	+	–	–
Shigella sonnei	ATCC #25931	+	–	–	–	–	–	+	–		–	–

(a) *Salmonella choleraesuis* ssp. *choleraesuis,* ser. typhimurium

2.3 Probe Selection

Oligonucleotides specific for target organism PCR products (Table 2-7) were designed and synthesized to confirm positive PCR results by probe hybridization as well as potentially increasing detection sensitivity. Probe confirmation is an important step in the application of molecular techniques for detection of organisms in environmental samples.

Technology transfer protocols, as well as quality control programs, should require the incorporation of a probe confirmation step, which can be achieved via conventional membrane hybridization of Southern transfers or directly blotted amplicons on a membrane, hybridization in a microplate format, by molecular beacons during PCR, or by microarray technology. In this project, a probe confirmation step was conducted and optimized for each target amplicon by conventional membrane hybridization and for a selected set of organisms by molecular beacons.

Table 2-7. PCR Probes Specific for Target Organism Genes.

Organism	Probe	Probe Sequences	Target Gene	Reference
Escherichia coli 0157:H7	SLTII-P	CACCGGGCAGTTATTTGC	shiga-like toxin gene	This report
Salmonella typhimurium	SHIMA-P	GGTAACTTCGGTCTGCGTGATAAAA	himA	Bej et al., 1994
Yersinia enterocolitica	YBHA-P	GTTTGCTTATACTCATCAGGG	ail gene	This report
Aeromonas hydrophila	AH16S-P	ATTTGGAGGCTGTGTCCTTG	16S rRNA	This report
Campylobacter jejuni	CJPG-P	GATGGCGATAGCAGATAGTT	flaA	This report
Helicobacter pylori	UREC-P	CCGGCGATGGCTTGGTGTGCGC	ureC gene	This report
Mycobacteriu m avium	MA-P	GGTCTCCGAGGCGCACAACG	dnaJ	Takewaki et al., 1993
Mycobacteriu m intracellulare	MI-P	GGACGCCAACCCGGACAATC	dnaJ	Takewaki et al., 1993
Shigella sonnei	SH-AP-P	CCATCTATTAGAATACCTGTG	ial locus	Schoolnik, 1993
Clostridium perfringens	PLC-P	TCAAAAGAATATGCAAGAGGT	phospholipase C gene	This report
MS2 Phage	MS2REP-P	AATTTAGGCTCGTTGTAGGG	MS2CG ssRNA	This report

2.4 Combining the GF/F-DE Filtration System with Molecular Detection Methods

This project evaluated two approaches for combining molecular detection methods with the GF/F-DE filtration method. The first approach involved extracting DNA directly from the concentrate resulting from 10-L filtrations of untreated source water. Amplifiable DNA was obtained from these extractions, but the scale of the extraction was too large to be considered practical for routine purposes. The nonsoluble nature of the GF/F-DE matrix necessitated large reagent volumes to be used in the extraction; over 100 mL of phenol/chloroform was necessary for recovery of DNA from a single 10-L filtration. In addition, DNA yields were low, possibly because DNA released from microbial cells was binding to the glass fibers and the silica particles comprising diatomaceous earth. Also, high concentrations of PCR inhibitors were co-extracted with the nucleic acids, requiring dilution of DNA samples for successful PCR detection. Despite the reduced sensitivity resulting from sample dilution, *Salmonella* spp.-specific DNA was detected in unseeded source waters using this approach. Nevertheless, the method was too unwieldy for practical purposes.

The second approach involved extracting DNA from a concentrate of cells following nutrient enrichment. A combination of selective or nonselective enrichment of test samples with molecular detection methods such as PCR is commonplace in the food industry. It allows relatively large sample volumes to be analyzed, increases the concentration of target organisms, and takes advantage of the specificity, sensitivity, and speed of molecular methods for detection. The use of a nonselective enrichment potentially permits the simultaneous detection of a number of pathogens, while selective enrichment may substantially increase the probability of detecting a specific target pathogen. This approach could also potentially provide valuable information about the ability or inability of culture-based methods to recover target organisms from complex microbial mixtures, since it is generally accepted that only 0.1 – 10% of bacteria in any particular environmental niche can be cultured on agar plates. Following filtration, the approach employed in this project involved incubating the entire GF/F-DE matrix in trypticase soy broth or selective media for various time periods, concentrating the resulting bacterial suspension by centrifugation, DNA extraction from the pellet, and detection of specific bacterial targets by PCR.

2.5 Evaluation of Concentrate Enrichment Combined with PCR

Studies were conducted to evaluate the efficacy of cultural enrichment combined with PCR for the detection of low concentrations of target organisms seeded into surface water samples. The basic procedure for the initial studies was as follows: 10-L volumes of untreated surface water were transported to the laboratory, seeded with target organisms and filtered through the 142-mm GF/F-DE composite filters as described in section 2.1.1. Following filtration, the filter-DE matrix was placed in 200 mL of either selective or nonselective enrichment broth and incubated for 0 to 24 hours. The liquid portion was carefully aspirated with a pipet into 50-mL Corex centrifuge tubes, avoiding carryover of the settled DE, and centrifuged at 10,000 rpm for 10 minutes. The supernatant was decanted, and the bacterial pellet was resuspended in phosphate buffer and centrifuged at 10,000 rpm for 10 minutes. After decanting the supernatant, the washed pellet was resuspended in Tris-EDTA, transferred to a polypropylene tube, and brought to a final volume of 5 mL prior to extraction of nucleic acids.

Many methods for extracting nucleic acids from environmental samples have been published, but no single technique has been developed which can be universally used to extract DNA or RNA from all types of environmental samples (see reviews by Tsai and Rochelle, 2001; Trevors and van Elsas, 1995). The most common methods used to physically disrupt and lyse bacterial cells in environmental samples include freeze-thawing, bead-beating, ultrasonication, and grinding. In addition to the physical disruption of the bacterial cells, lytic enzymes such as lysozyme, proteinase K, and pronase have been commonly used to breakdown bacterial cell structures. Detergents such as SDS and sarkosyl are typically incorporated enzymes to accelerate the lysis process, while the addition of EDTA to the lysis buffer reduces or eliminates the activity of nucleases. An alternative to enzymatic treatments has been the addition of a strong chaotropic agent, guanidine thiocyanate, to lyse bacterial cells and preserve released DNA or RNA. The cation-exchange resin Chelex-100 has been demonstrated to disperse soil particles and bind some inhibitory compounds, thus aiding recovery of nucleic acids from environmental samples. A mixture of phenol and chloroform is commonly included in extraction protocols to remove proteins and lipids from environmental extracts. The procedure selected for the initial development of methods for this project (Figure 2-3) has been shown to be widely applicable to a variety of environmental sample types.

1. Add 18 μL of 18 mg/mL proteinase K and 2 μL 10% (w/v) SDS to 400 μL of enrichment concentrate, vortex, and incubate at 37°C for 30 min.
2. Add 200 μL 2% (w/v) sarkosyl, vortex, and incubate at 37°C for 15 min.
3. Add 175 μL 5 M NaCl, vortex, and freeze (-70°C for 10 min)/thaw (65°C for 5 min), three times. Frozen samples can be stored at -80°C.
4. Add 170 μL of 5% hexadecyltrimethylammonium bromide (CTAB) in 1 M NaCl, vortex for 1 min, and incubate at 65°C for 30 min.
5. Add 865 μL of phenol/chloroform/isoamyl alcohol (25/24/1) and shake sample tubes to homogenize suspension.
6. Centrifuge at 10,000 x g for 5 min at 20°C.
7. Transfer 760 μL supernatant to 1.5-ml centrifuge tubes and add 760 μL chloroform/ isoamyl alcohol (24/1), vortex, centrifuge at 10,000 x g for 5 min at 20°C.
8. Transfer 700 μL supernatant to 1.5-ml centrifuge tubes, add 1 μL glycogen, mix by inversion, and incubate at room temperature for 5 min.
9. Add 420 μL isopropanol, mix gently by inversion, and store overnight at -20°C.
10. Centrifuge at 10,000 x g for 15 min at 4°C.
11. Carefully discard supernatant, and add 500 μL ice-cold 70% molecular grade ethanol, and centrifuge at 10,000 x g for 15 min at 4°C (Repeat this step).
12. Carefully discard the supernatant and air dry the DNA pellets.
13. Resuspend DNA pellets in 200 μL sterile PCR-grade water and allow to dissolve for 30 min at room temperature.

Figure 2-3. Procedure for Initial Development of Methods.

Glycogen was added as a co-precipitant to aid in the recovery of low concentrations of DNA. However, the effect of glycogen precipitation on subsequent PCR analyses has not been extensively evaluated. Therefore, DNA extracted from unseeded source water samples was precipitated using the same concentration of glycogen applied in the protocol above and amplified using primers specific for the *lacZ* gene (a coliform marker). In addition, glycogen was added directly to PCR reactions at concentrations in the range 0.1 to 1 μg/μL. Glycogen did not produce any inhibitory affect in any of these reactions.

To evaluate the utility of the combined enrichment-PCR approach, 10 L of untreated surface water was seeded with 100 CFU of *Salmonella cholerasuis*, ser. typhimurium and concentrated using the GF/F-DE system. The filter was selectively enriched with SBG broth for incubation periods of zero, two, six, and 24 hours. An aliquot of the enrichment broth was removed for culture-based isolation at each time period, and the remaining enrichment broth was used for DNA extraction and PCR-based detection. *S. typhimurium* was detected by the cultural method at all enrichment time periods, including the zero time sample that had enrichment medium added but was not incubated. Detection in the zero time sample demonstrated the efficacy of the concentration procedure and also indicated the prevalence of *Salmonella* spp. in untreated surface waters. PCR amplification was performed using the SHIMA primer set on dilutions of the extracted DNA, followed by hybridization and probe confirmation (SHIMA-P probe). As with the cultural method, the presence of *Salmonella* spp.-specific target DNA was detected in all samples from the different incubation periods.

Although selective enrichment of target organisms should greatly increase the concentration of target cells and hence the number of copies of target DNA, other organisms that are present but unable to compete with the more rapidly growing bacteria in an enriched sample may not be detected. Therefore, further experiments evaluated the use of trypticase soy broth (TSB) as a nonselective means of enriching a broad range of potential target organisms. A single nonselective enrichment of a filtered sample followed by a single DNA extraction allows for subsequent PCR-based detection of multiple target organisms, rather than relying on multiple, selective enrichments for each organism of interest. Experiments were conducted as described above except that 10-L surface water samples were simultaneously seeded with three target organsims (*S. cholerasuis*, ser. typhimurium, *E. coli* 0157:H7, and *Y. enterocolitica*) at concentrations of 10 CFU/10 L and 100 CFU/10 L. Following addition of TSB, samples were incubated for zero, two, six, and 24 hours. Culture-based detection by plating was not conducted for this experiment. None of the target organisms were detected by PCR during this experiment at any of the incubation periods, possibly due to a combination of interference and inhibition (see Section 2.6). It was likely that nonselective enrichment of the large volume sample increased the concentration of nontarget cells relative to the three target organisms resulting in masking of the target DNA in subsequent molecular analyses.

As an alternative to the extraction protocol described above, two commercially available DNA extraction kits were evaluated for the capacity to yield amplifiable DNA from the enrichment pellets. Previous work at MWDSC demonstrated that a DNA extraction kit designed for soil (FastDNA SPIN Kit for Soil, Bio101, Carlsbad, Calif.) was useful for the isolation of whole community DNA from bacteria associated with carbon filtration media particles from a pilot-scale water treatment facility (Burr et al., 2000). The basic concept of this kit was mechanical shearing (bead-beating) in the presence of a lysis reagent to disrupt cellular membranes and release nucleic acids, and subsequent purification of the DNA. Use of these kits provided PCR-quality DNA from samples containing high concentrations of substances inhibitory to PCR, so it was therefore likely that these kits would yield amplifiable DNA from the enrichment matrix. A second kit, also designed for extracting DNA from soil, was also based on a bead-beating method but included a proprietary inhibitor removal solution (MoBio Laboratories, Inc., Solana Beach, Calif.). A comparison of the two kits indicated that the yield of DNA obtained from latter kit was superior and the extracted DNA was suitable for PCR-based detection using the range of primer sets selected for this study without dilution or further purification. Consequently, the majority of enrichment-extractions were carried out using the MoBio kit using the method outlined in Figure 2-4 (many of the reagents are proprietary and thus are referred to as per the manufacturer's label). The advantages of such a kit include the commercial availability of a complete kit, avoidance of toxic compounds such as phenol and chloroform, and less variation between reagent lots (so long as the manufacturer has good quality control practices in place). A recent study described the use of this kit for DNA extraction from soil and reported that 50% of samples yielded amplifiable DNA (Rosch et al., 2002).

1. Add 300 µl of resuspended, washed enrichment pellet to Bead Solution tube and gently vortex to mix.
2. Add 60 µl of Solution S1, invert once to mix, and add 200 µl of Inhibitor Removal Solution (IRS).
3. Secure tubes horizontally to flat-bed pad, and vortex at maximum speed for 10 min.
4. Centrifuge for 30 sec at 10,000 x g and transfer 725 µl of supernatant to a clean microfuge tube.
5. Add 250 µl of Solution S2, vortex, incubate at 4°C for 5 min, and centrifuge tubes for 1 min at 10,000 x g.
6. Transfer 450 µl of the supernatant to a clean microfuge tube, carefully avoiding the pellet.
7. Add 900 µl of solution S3 to the supernatant, and mix well.
9. Discard the filtrate, add remaining supernatant to the SPIN filter, and centrifuge for 1 min at 10,000 x g (2 centrifugations are required for each sample).
10. Add 300 µl of Solution S4, and centrifuge for 30 sec at 10,000 x g.
11. Discard the filtrate, and centrifuge again for 1 min.
12. Carefully place SPIN filter in a new tube (avoid splashing any Solution S4 onto the SPIN filter).
13. Add 50 µl of PCR water to the center of the white filter membrane, and centrifuge for 30 sec at 10,000 x g.
14. Discard SPIN filter and combine duplicate tubes (final volume of 100 µl per sample).

Figure 2-4. Enrichment-Extraction Procedure Using the MoBio Kit.

A series of experiments was performed to evaluate the soil extraction kit with 10-L volumes of surface water seeded with *S. cholerasuis*, ser. typhimurium, *E. coli* 0157:H7, and *Y. enterocolitica*. The 10-L samples were concentrated by the GF/F-DE method, and the filters were incubated in TSB overnight. Although both the cultural and PCR-based methods detected target organisms with very short enrichment incubation periods, an overnight incubation was selected based on the practicality of collecting and concentrating a sample on one day, enriching overnight, and then extracting DNA and performing PCR the next day, providing pathogen detection within 24 hours. Incubation temperatures of 25°C and 35°C were used for the enrichment, since 25°C may provide *Y. enterocolitica* with a more competitive advantage relative to other, faster-growing heterotrophs in the sample. A small aliquot of the enrichment culture was removed for plating; the remaining enrichment broth was extracted using the soil extraction kit and the resulting DNA used for PCR-based detection using the SHIMA, SLTII, and BHA primer sets followed by probe confirmation. *Salmonella* was recovered only once from the TSB nonselective enrichment using conventional plating methods (Table 2-8). *E. coli* O157:H7 and *Y. enterocolitica* were not recovered by culture methods from samples seeded with up to 100 CFU/10 L. This was not surprising since the nonselective enrichment medium would potentially increase the populations of all heterotrophs present, making culture-based isolation more difficult. In contrast, DNA corresponding to *Salmonella* spp. and toxigenic *E. coli* was detected in all of the overnight enrichment samples when using undiluted DNA as the template (Table 2-8). In addition, *Salmonella* spp.-specific DNA was detected in two out of the three unseeded water samples. These data confirm the utility of the enrichment/PCR approach for detecting low concentrations of pathogens in water (Figure 2-5).

Table 2-8. Detection of Seed Organisms in 10-L Surface Water Samples after 24 Hours Nonselective Enrichment.

Organism	Trial	Seed conc. CFU/10-L	Background PCR[a]	Presence/Absence following enrichment at: 25° C PCR	25° C Culture	35°C PCR	35°C Culture
S. typhimurium	1	100	–	+	–	+	–
S. typhimurium	2	100	+	+	–	+	+
S. typhimurium	3	10	+	+	–	+	–
E.coli 0157:H7	1	100	–	+	–	+	–
E.coli 0157:H7	2	100	–	+	–	+	–
E.coli 0157:H7	3	10	–	+	–	+	–
Y.enterocolitica	1	100	–	–	–	–	–
Y.enterocolitica	2	100	–	–	–	–	–
Y.enterocolitica	3	10	–	–	–	–	–

(a) Unseeded 10-L surface water sample concentrated and enriched for 24 hours at 35°C and analyzed by PCR.

Enrichment at 25° or 35°C did not result in detection of *Y. enterocolitica* during PCR amplification using the BHA primer set, which may be due to the nonselective conditions present. It appears that the nonselective enrichment procedure for *Y. enterocolitica* does not improve detection of this organism. Given the success of the first combined enrichment experiment using selective media for *Salmonella,* the use of specific selective enrichment for *Yersinia* in conjunction with PCR detection may provide a viable option for more rapid detection of yersiniae in environmental samples; however, that option was not explored in this project. Based on the successful PCR detection of low concentrations of *Salmonella* and *E. coli* 0157:H7 from 10-L concentrates, the combined enrichment/PCR method was adopted for further evaluation during the field-sampling portion of the project.

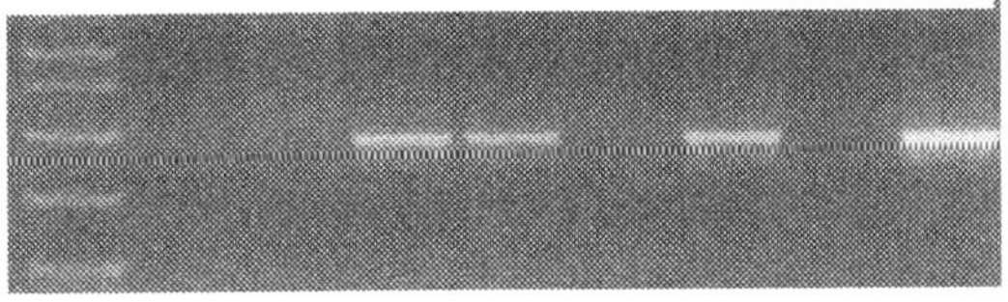

Figure 2-5. Detecting *Escherichia coli* O157:H7 Using Combined Enrichment/PCR Approach.
Amplicons (482 bp) were obtained from 10-L source water samples seeded with 10 CFU of *E. coli* and enriched at 25°C (Lanes 4 and 5) or 37°C (Lane 7). Lanes 2, 3, and 6: unseeded source water; Lanes 8 and 9, negative and positive controls, respectively; Lane 1, molecular weight marker.

2.6 Evaluation of Membrane Dissolution Concentration Method

An alternative approach to the application of molecular methods for detection of target organisms involved filtering water samples through polycarbonate membrane filters (47- to 142-mm diameter) followed by extraction of DNA from entrapped microorganisms directly on the filter. This extraction takes advantage of the solubility of polycarbonate in phenol/chloroform and avoids the problems associated with removing bacteria from the filter prior to extraction. Preliminary experiments investigated the use of 47-mm, 0.2-μm porosity polycarbonate filter membranes to concentrate 100-mL samples of untreated source water. DNA was extracted using an adaptation of the CTAB extraction process described above. Following filtration, the membranes were folded and immersed in 400 μL of 50 mM Tris-HCl/20 mM EDTA, pH 8, in a 1.7-mL microfuge tube. The membranes were then incubated with 100 μg/ml proteinase-K for 30 minutes at 37°C, followed by the addition of sarcosine (1% final concentration) and incubation for a further 15 minutes. Sodium chloride was added to a final concentration of 1.1 M, and the samples were then subjected to three cycles of freezing (-70°C) and thawing (65°C). CTAB was added to a final concentration of 1%, and the samples were incubated at 65°C for 30 minutes. An equal volume (750 μl) of phenol/chloroform/isoamyl alcohol (25/24/1) was added, and following rapid emulsification to dissolve the filter, the samples were centrifuged at 12,000 x g for 15 minutes at 4°C. DNA was precipitated from the aqueous phase by the addition of 600 μl of isopropanol and incubation overnight at -20°C, followed by centrifugation at 12,000 x g for 15 minutes at 4°C. The DNA pellet was washed with 1.0 ml of ice-cold 70% ethanol, centrifuged at 12,000 x g for 15 min at 4°C, dried in a desiccator for 20 min, and finally resuspended in 100 μl of sterile distilled water.

Gel electrophoresis of the total DNA extracted from these samples indicated large quantities of high-quality undegraded DNA. The general suitability for PCR of DNA extracted using this method was assessed using universal primers. Twenty-four unseeded environmental water samples (100 mL) were concentrated and extracted using the membrane dissolution procedure. Extracted DNA was amplified with primers targeting conserved regions of the bacterial 16S rRNA gene that should produce a 320 bp amplification product with over 95% of bacterial species. Fifteen of these samples (65%) generated amplifiable DNA. Amplification of DNA extracted from water samples seeded with *E. coli* 0157:H7 and *S. typhimurium* using the SLTII and SHIMA primer sets, respectively, was successful, indicating that the quality of the

DNA was sufficient for specific PCR detection. The sensitivity of the membrane dissolution procedure, based on seeding serial dilutions of bacteria into treated drinking water, was 1 CFU/100 mL for *E. coli* O157:H7 and 10 CFU/100 mL for *Y. enterocolitica.* The sensitivity of the procedure was also evaluated with source water. Dilutions of an overnight culture of *E. coli* O157:H7 (enumerated by a standard plate count on nutrient agar) were seeded into 100 mL of untreated source water (1.5 NTU). Using primers targeting the stx2 gene, 5 CFU/100 mL were readily detected. No stx2-specific DNA was detected in unseeded source water samples. A similar experiment with *S. typhimurium* demonstrated that 3 CFU/100 mL could be detected. *Salmonella* spp.-specific DNA was detected with the SHIMA primers in most unseeded, untreated source water samples. A summary of the evaluation of the membrane dissolution extraction with a variety of pathogens and formats is provided in Table 2-9.

Table 2-9. Detection Sensitivities for Membrane Dissolution-PCR Procedure During Method Evaluation Phase.

Seed conc.	Proportion (%) of positive PCR reactions with primers specific for: [a]		
(CFU/matrix)	*E. coli* O157	*S. typhimurium*	*Y. enterocolitica*
100 mL source water			
Unseeded	0 (4) [b]	80 (5)	nd
1–5	100 (8)	75 (8)	nd
5–30	100 (4)	100 (4)	nd
>30	100 (5)	100 (9)	nd
100 mL - 1 L treated effluent			
Unseeded	0 (5)	100 (4)	0 (4)
1–5	80 (15)	100 (8)	71 (7)
5–30	100 (5)	100 (4)	100 (4)
>30	100 (6)	100 (6)	100 (7)
1 L source water			
Unseeded	0 (2)	66 (3)	0 (2)
1– 5	50 (4)	83 (6)	60 (5)
5–30	100 (2)	100 (3)	100 (1)
>30	100 (2)	100 (3)	nd
10 L treated effluent			
Unseeded	nd	0 (2)	0 (2)
1– 5	nd	100 (4)	50 (4)
5– 30	nd	100 (2)	100 (2)

(a) Using the stx2, SHIMA, and BHA primers.
(b) Numbers in parentheses are the number of reactions performed. Nd = Not determined.

A disadvantage of this extraction technique is the use of hazardous reagents (phenol and chloroform). Therefore, an alternative DNA extraction procedure was evaluated which substituted a nontoxic compound, ProCipitate (CPG Inc., Lincoln Park, N.J.) for the phenol step in the standard extraction procedure. ProCipitate is a synthetic, water-insoluble protein-binding alternative to phenol for extraction of nucleic acids. It has been used previously in the extraction of viral nucleic acids from environmental samples as part of a PCR-based detection assay. This method produced good quality DNA, but the yield from seeded water samples was about 50% compared to the phenol-based extraction, probably due to the inability of the ProCipitate-based method to dissolve the polycarbonate membrane. Much of the extracted DNA likely remained associated with the membrane when removed from the extraction. Since detection sensitivity was a primary issue in this project, the inconvenience of working with phenol was considered more acceptable than the loss of sensitivity likely with the less toxic extraction method.

Other methods of membrane dissolution DNA extraction evaluated included using larger diameter membranes and self-contained cartridge filters. Increasing the diameter of the 0.22-μm porosity membrane filters from 47 mm to 90 mm and 142 mm greatly increased the volume of raw and treated drinking water that could be filtered, allowing for potential increases in detection sensitivity. However, increasing the volume that could be filtered was highly dependent on the sample matrix; while 10 L of treated drinking water could be filtered through 142-mm membranes in less than one hour, raw water sample volumes above 5 L required more than two hours. In addition, using larger diameter membranes required scaled-up versions of the standard DNA extraction procedure, thereby increasing the volumes of reagents by up to four-fold to ensure complete dissolution of the membranes.

In order to keep the method of membrane dissolution concentration practical in terms of available materials and methods, 0.22-μm porosity, 47-mm polycarbonate membranes were selected for use during the survey phase of the project. While the volume of sample that could be easily filtered through the 47-mm filters was considerably less than that using the larger diameter filters, the methods and materials involved are common to most analytical laboratories specializing in water or wastewater. Self-contained cartridge filters have been used previously for the direct isolation of nucleic acids from seawater (Sommerville et al., 1989) and may allow large volumes to be filtered while minimizing the volume of extraction reagents required. Commercially available 0.45-μm porosity filter cartridges (Sterivex-GS; Millipore Corp., Bedford, Mass.) were used in combination with a dissolution/extraction protocol to filter up to 5 L of raw water, and the DNA was extracted using a sucrose/EDTA/Tris buffer wash, followed by a lysozyme/proteinase-K extraction as described previously (Sommerville et al., 1989). However, no amplifiable DNA was recovered from any of the preliminary experiments using this approach. Further evaluation of this alternative was terminated, as it was somewhat difficult to adapt the cartridges to existing laboratory processes without complicated modifications, thereby reducing the practicality for this project.

2.7 Evaluation of Molecular Beacons

As described in the introduction, molecular beacon technology is one of a number of approaches for performing PCR-based detection assays in a rapid homogeneous format. The conventional PCR-based approach to pathogen detection involves PCR amplification of the desired target, analysis of amplification products by agarose gel electrophoresis, transfer of amplified products to a nylon membrane, and finally confirmation of amplicon identity by hybridization with a specific probe internal to the PCR primers. In contrast, the molecular beacon approach incorporates PCR amplification and probe hybridization into a single assay and the automated acquisition of assay results (detection of amplified product) in real time as the PCR assay progresses. The term "real-time" as it is applied to describe molecular beacon assays and other similar quantitative PCR technologies means that the analyst can detect the assay products (amplified DNA) while the amplification is still in progress. It does not provide real-time detection of pathogens in water. Water samples still have to be collected and concentrated, and the nucleic acids extracted in a form suitable for amplification. Consequently, any sample concentration and extraction methods that are compatible with conventional PCR methods will also be suitable for molecular beacon assays or other quantitative PCR assays. Similarly, any inhibitory compounds in environmental samples that are co-extracted with the nucleic acids may result in false-negative results with molecular beacon assays just as they would for conventional PCR methods. Also, the selection, evaluation, and optimization of primer sequences are just as important for molecular beacon assays as for conventional PCR amplification. In fact, it may be more important because the primers and molecular beacon probe are included in the same reaction tube, leading to possible interferences between the primers and probes if sequences are not selected carefully. The following section describes the development and evaluation of molecular beacon assays for *Salmonella* spp. and *E. coli* O157:H7. All assay sensitivities should be regarded as nominal because they were obtained by making serial dilutions of extracted DNA.

2.7.1 Selection of PCR Primers and Probes Used for Molecular Beacon Assays

Primers were synthesized by Genosys and molecular beacon probes were synthesized by Midland Certified Reagent Company (Midland, Tex.). The sequences of all primers and probes used in the development of molecular beacon assays are provided in Table 2-10. Primers were resuspended in TE buffer and stored at -20°C, and 20-μM aliquots of each primer were prepared and used for PCR. Probes were stored –20°C as 20-μM aliquots in 20 mM Tris-HCl, pH 8.0, 1 mM $MgCl_2$.

Table 2-10. Primer and Probe Sequences Used for Molecular Beacon Assays.

Primer/ probe	Sequence (5' - 3')	Reference
Escherichia coli O157:H7		
O157AF	AAGATTGCGCTGAAGCCTTTG	Desmarchelier et al., 1998
O157AR	CATTGGCATCGTGTGGACAG	
O157BF	AAATATAAAGGTAAATATGTGGGAACATTTGG	This project and Fortin et al., 2001
O157BR	TGGCCTTTAAAATGTAAACAACGGTCAT	
O157rfbe[a]	CGCTATGGTGAAGGTGGAATGGTTGTCACGAATAGCG	This project and Fortin et al., 2001
Salmonella spp.		
SHIMAL	CGTGCTCTGGAAAACGGTGAG	Bej et al., 1994
SHIMAR	CGTGCTGTAATAGGAATATCTTCA	
BHMA1[a]	CGCTATCCGGGGCGTAACCCGTAGCG	This project and Chen et al., 2001
Escherichia coli		
EHIMAF	CGCGCTCTGGAAAACGG	This project and Chen et al., 2000
EHIMAR	CGTGCTGTAATGGGAATATC	

[a] Underlined sequences correspond to the complementary sequences that form the stem structure in nonhybridized molecular beacon probes. Probes were labeled with fluorescein (6-FAM) at the 5' end and 4-(4'-dimethylaminophyenylazo) benzoic acid (DABCYL) at the 3' end.

The O-antigen gene cluster (*rfb* operon), encoding enzymes necessary for biosynthesis of the O-antigen, was used as the target for the development of a real-time, quantitative PCR assay for *E. coli* O157:H7. This cluster of genes is highly conserved among strains within the *E. coli* O157 serovar. For development of the molecular beacon assay, primers and probes were targeted within the *rfb*E gene (Figure 2-6). This is a different region of the O-antigen operon than that targeted for the conventional *E. coli* O157 PCR assay described earlier in this chapter and used for the pathogen survey. Primer pair O157AF and O157AR (496 bp amplicon) was described previously for detection of *E. coli* O157 by conventional PCR (Desmarchelier et al., 1998). A second set of primers (O157BF and O157BR) was designed to amplify a smaller 145 bp fragment of the same *rfbE* region. The 37 bp molecular beacon (O157rfbe) contained a 25 bp region that was designed to perfectly match the *rfbE* gene of *E. coli* O157:H7, flanked by two 6 bp arms that hybridize to each other in the presence of $MgCl_2$ when no complementary target sequence is present. The 5' end of the probe was labeled with fluorescein (6-FAM) as the reporter molecule (excitation at 418 nm and fluorescence emission at 518 nm) and 4-(4'-dimethylaminophenylazo) benzoic acid (DABCYL) was used as the quencher at the 3' end. This probe hybridized internally to both sets of *E. coli* O157:H7 primers.

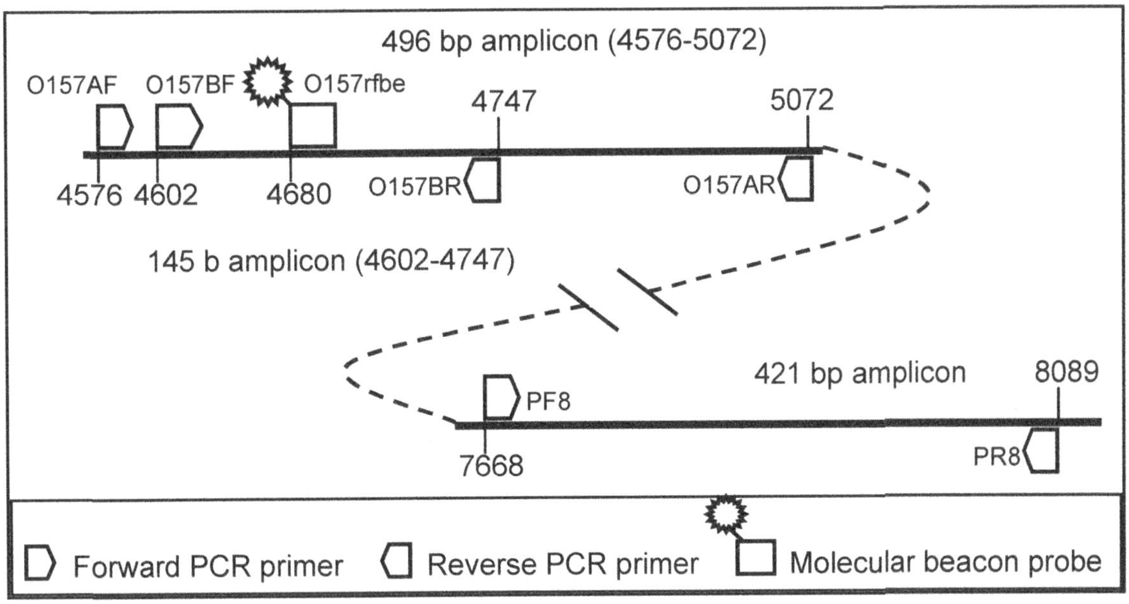

Figure 2-6. Schematic Representation of Fragment of *Escherichia coli* O157:H7 O-antigen Gene Cluster (*rfb* operon). Display the positions of all primers and probes used throughout this project. The numbers associated with each primer/probe correspond to the base position within the O-antigen gene cluster (GenBank accession number AF061251; Wang and Reeves, 1998) and the alphanumeric codes correspond to the names used in Table 2-5. The region spanned by primers O157AF and O157AR was within the *rfbE* gene, while the region spanned by PF8 and PR8 was within the *rfbB* gene.

The *himA* gene of *Salmonella* spp., encoding a DNA-binding protein, was used as the target for the real-time PCR assay with the same primers used for the conventional PCR assays described elsewhere in this project (SHIMAL and SHIMAR, 122 bp amplification product, Table 2-4). A similar pair of primers was also designed to amplify the corresponding *him*A region of *E. coli*. The molecular beacon (BHMA1) was designed to be perfectly complementary to the *himA* gene of *Salmonella* spp. but contained two nucleotide mismatches relative to the same region in the *himA* gene of *E. coli*.

2.7.2 Development of Molecular Beacon-Based Real-Time PCR Assay for Detection of *E. coli* O157:H7

Genomic DNA isolated from *E. coli* O157:H7 ATCC 43985 was used as the template (2.5 μL) in amplification reactions containing 27.5 μL of PCR master mix (3.5 mM $MgCl_2$, 0.5 μM of each primer, 0.2 μM of dNTPs mix, 1 μM of O157rfbe probe, 5 U of AmpliTaq Gold DNA polymerase, and 2.5 μL TaqMan buffer A, made up to 100 μL with water. Both the polymerase and buffer A were supplied as part of the TaqMan PCR core reagent kit (PE Biosystems, Foster City, Calif.). All reactions were performed in the 200-μL MicroAmp optical tubes sealed with the MicroAmp optical caps (PE Biosystems, Foster City, Calif.). The Perkin-Elmer ABI Prism 7700 sequence Detection System was used for real-time analyses.

Amplification reactions consisted of four steps. Samples were heated at 95°C for 10 minutes, followed by 40 cycles of melting at 94°C for 45 seconds, fluorescent measurement at 41°C for 30 seconds, annealing at 52°C for 45 seconds, and extension at 72°C for 45 seconds. Upon completion of 40 thermal cycles, data was automatically analyzed by the ABI Prism 7700 system, and amplification plots were obtained. The results of real-time quantitative PCR assays are typically expressed as values for the threshold cycle (C_t). This is the cycle number at which the fluorescence signal resulting from amplification of the target nucleic acid becomes clearly discernible from the background fluorescence levels. For the purposes of this study the C_t was defined as the first PCR cycle at which the fluorescence was 10-fold higher than the standard deviation of the mean baseline emission. The four-step thermal cycle profile with signal acquisition at 41°C and a primer annealing temperature of 52°C provided improved sensitivity over a three-step cycle with signal recording and primer annealing at 45°C. PCR product from the 496 bp fragment of the *rfbE* gene generated with primers O157AF and O157AR was detected after 30 cycles of amplification ($C_t = 30$, Figure 2-7). The smaller amplicon (146 bp, O157BF and O157BR) generated twice as much fluorescent signal upon completion of thermal cycling and was detected earlier ($C_t = 26$, Figure 2-7).

The sensitivity of the real-time PCR assay was evaluated using 10-fold serial dilutions of template DNA isolated from *E. coli* O157:H7 strain ATCC 43895 (Figure 2-8A). Using the C_t criterion of positive detection being a fluorescent signal 10-fold higher than the standard deviation of the mean baseline emission, the detection limit of the assay was determined to be 5×10^3 CFU/mL, a value comparable to that reported using other fluorogenic real-time PCR assays for *E. coli* O157:H7 (Sharma et al., 1999; Oberst et al., 1998). The sensitivity of the enrichment-conventional PCR approach for *E. coli* O157:H7 was 10 CFU, as described earlier in this report, using primers targeting the Stx2 gene (Table 2-8), and the original publication describing conventional PCR with the P8 primers targeting the *rfbB* gene reported a sensitivity of 10 cells (Maurer et al., 1999). Since the C_t value represents the PCR cycle at which the fluorescent intensity rises above the threshold, it can be used to quantify the input target concentration (Figure 2-8). This quantification proved to be linear over a wide range of initial target concentrations (from 10^3 to 10^8 CFU/mL).

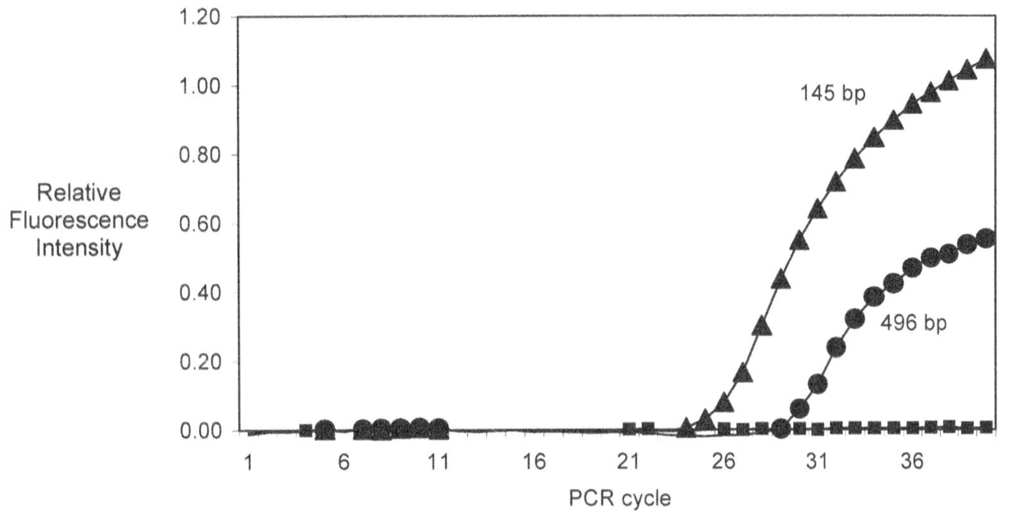

Figure 2-7. Normalized Fluorescence Measurement (ΔRn) as Function of Number of Amplification Cycles for Real-Time Detection of *Escherichia coli* O157:H7 (ATCC 43985).
When amplifying 145 bp (▲) and 496 bp (●) regions of the *rfbE* gene. ■, No template control.

2.7.3 Specificity of *E. coli* O157:H7 Molecular Beacon Assay

The specificity of the MB-based real-time PCR assay relies on the selected sequence of the primer set and the probe moiety of the beacon, both of which were based on the *rfbE* gene of *E. coli* O157:H7 strain 86-24. Genomic DNA recovered from 28 shiga toxin-producing *E. coli* (STEC) strains and seven other non-STEC strains was amplified to assess the ability of the real-time PCR assay to discriminate the O157 serotype from non-O157 strains (Table 2-11). All of the 11 reference strains of *E. coli* serotype O157:H7 generated positive results with the molecular beacon assay. In addition, four non-H7 strains harboring the somatic O157 antigen were also positively detected. In contrast, the remaining non-O157 STEC strains did not emit fluorescence upon PCR amplification, regardless of their genetic profiles. This covered 11 different serovars including *E. coli* 055:H7, a serotype of enterohemorrhagic *E. coli* that is closely related to the O157:H7 serovar. None of the non-STEC strains were detected by the real-time PCR assay. The specificity of the molecular beacon assay was also tested against several bacteria known to cross-react with the O157 antisera, including *Escherichia hermanii, Citrobacter freundii, Vibrio cholerae, Y. enterocolitica,* and *Salmonella* group N. None of these bacteria produced a detectable fluorescence signal (Table 2-12). Although a similar *rfb* gene cluster has been cloned from *V. cholerae*, *Y. enterocolitica*, and *Salmonella* spp. (Al-Hendy et al., 1991; Brown et al., 1992; Morona et al., 1990), none of these bacteria were detected.

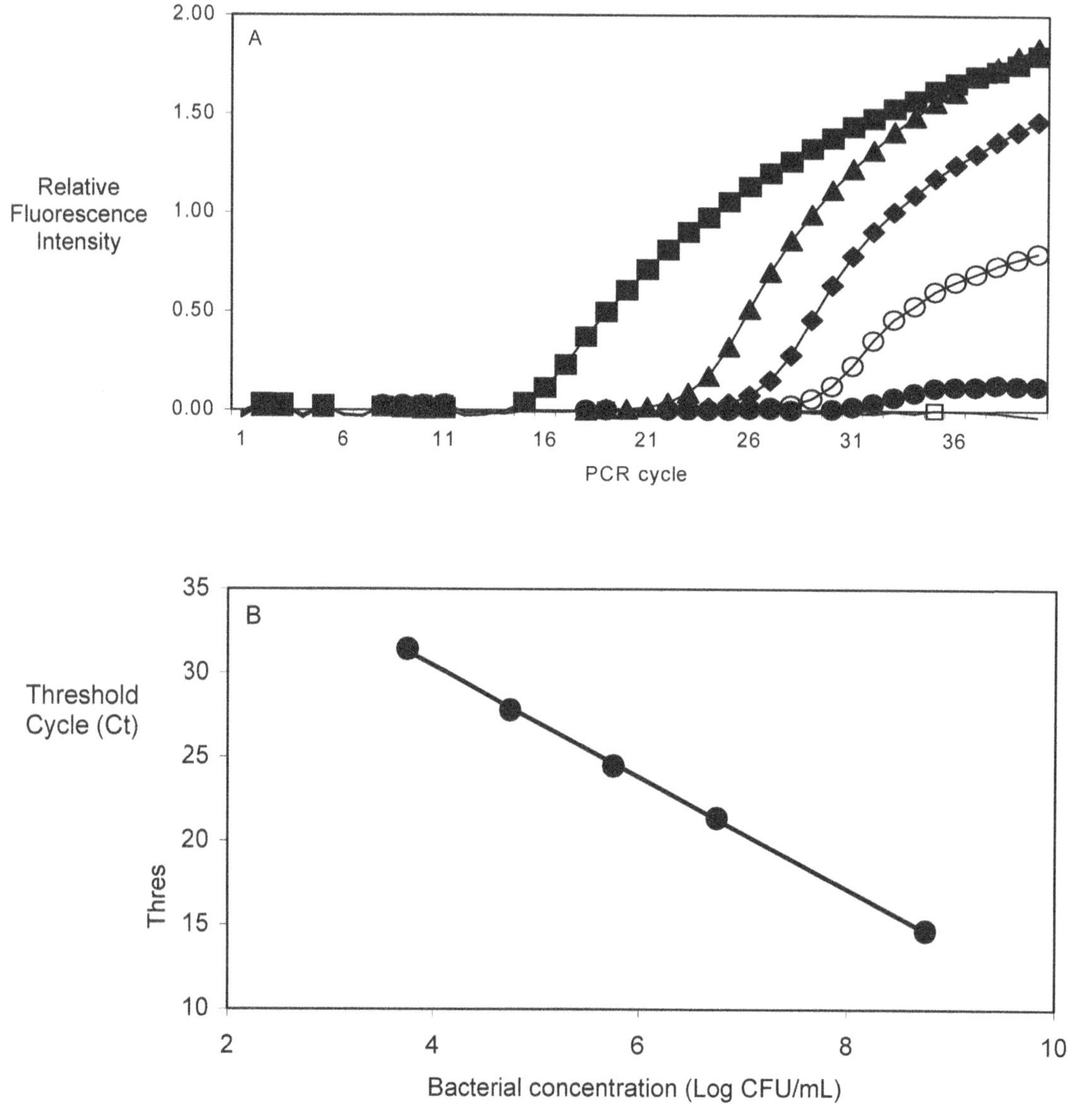

Figure 2-8. Sensitivity (A) and Standard Curve (B) of *Escherichia coli* O157:H7-Specific Molecular Beacon Assay Based on Serial 10-fold Dilutions of DNA Template Extracted from *E. coli* O157:H7 ATCC 43985.
Initial bacterial concentrations in (A) were: No template control (□), 5×10^3 (●), 5×10^4 (○), 5×10^5 (♦), 5×10^6 (▲), and 5×10^8 CFU/mL (■).

Table 2.11. *E. coli* Serotypes Tested for Specificity of Molecular Beacon Assay.

Bacterial Strain	Serotype	Genotype[(a)]	Detection[(b)]	Source[(c)]
E. coli ATCC 35150	O157:H7	*stx1, stx2*	+	ATCC
E. coli 4718	O157:H7	*stx1, stx2, eaeA*	+	1
E. coli 2725	O157:H7	*stx1, stx2, eaeA*	+	1
E. coli 2409	O157:H7	*stx1, eaeA*	+	1
E. coli 6363	O157:H7	*stx1*	+	2
E. coli ATCC 43895	O157:H7	Unknown	+	ATCC
E. coli	O157:H7	Unknown	+	
E. coli 1880	O157:H7	*eaeA*	+	3
E. coli 2977	O157:H7	*eaeA*	+	3
E. coli 5906	O157:H7	*eaeA, stx2*	+	3
E. coli B6-914	O157:H7	*eaeA*	+	3
E. coli 3048	O157:NM	*stx1, stx2, eaeA*	+	1
E. coli MF7123A	O157:NM	None	+	3
E. coli MF13180	O157:NM	*eaeA, stx1, stx2*	+	3
E. coli MF13180-25	O157:NM	*eaeA, stx1, stx2*	+	3
E. coli	O157:H⁻	*stx1, stx2, eaeA*	+	3
E. coli DEC5A	O55:H7	Unknown	-	1
E. coli	O5:NM	*stx1*	-	3
E. coli 5710	O26:H11	*stx1, stx2, eaeA*	-	1
E. coli 13C60	O26:H11	*stx1*	-	2
E. coli 90.0105	O26:H11	*stx1*	-	3
E. coli 3128	O113	*stx1, stx2*	-	1
E. coli 3883	O111:NM	*stx1, eaeA*	-	1
E. coli 403	O111:NM	*stx1*	-	2
E. coli 13C62	O103:H2	*stx1*	-	2
E. coli 5702	O103:H2	*stx1*	-	1
E. coli 87.1468	O103:H2	*stx1*	-	3
E. coli 15597	Unknown	Unknown	-	4
E. coli Seff a	Unknown	Unknown	-	
E. coli Seff b	Unknown	Unknown	-	
E. coli FAMP	Unknown	Unknown	-	4
E. coli C	Unknown	Unknown	-	4
E. coli CN13	Unknown	Unknown	-	4
E. coli W3110	Unknown	Unknown	-	

[(a)] The presence of the virulent genes was indicated by the donor/source.
[(b)] + or - indicates that the relative fluorescent intensity (ΔRn value) after 40 cycles was greater than the threshold value, calculated at the 95% confidence level. Each PCR reaction was initiated with ~10^7 DNA molecules.
[(c)] Isolates were supplied by: [1]ARS, USDA, Ames, IA ; [2]CFSAN, USFDA, Washington, D.C. ; [3]ARS, USDA, Philadelphia, PA.; CSDLAC, Whittier, CA.

Table 2-12. Specificity of *E. coli* O157 Molecular Beacon Assay with Strains Known to Cross-React with O157 Antisera.

Bacteria	Cross- reaction	Detection[(a)]	Source[(b)]
E. hermannii strain 3390-90	+	-	1
E. hermannii strain 1000-80	+	-	1
E. hermannii strain 3018-85	-	-	1
E. hermannii strain 3607-89	-	-	1
C. freudii strain F90/2460-1	+	-	1
Salmonella strain 425-82	+	-	1
Salmonella strain 1552-82	+	-	1
Salmonella LT2	unknown	-	
V. cholerae Pacini O:1 ATCC 11623	+	-	ATCC
Yersinia enterocolitica ATCC 23715	unknown	-	ATCC
Enterobacter cloacae ATCC 13040	unknown	-	ATCC
Citrobacter freundii ATCC 8090	unknown	-	ATCC

(a) + or - indicates that the relative fluorescent intensity (ΔRn value) after 40 cycles was greater than the threshold value, calculated at the 95% confidence level.
(b) Isolates were supplied by: [1]Dr. N. Strockbine, CDC, Atlanta, Ga.

2.7.4 Application of Molecular Beacon Assay to Environmental Water Samples

Of six environmental water samples that were positive with the stx2-specific primer pair, only one was positive with the *E. coli* O157:H7-specific (*rfbE* gene) molecular beacon assay (Table 2-13). The stx2 primers/probe combination target a shiga toxin-like gene that is present in many bacteria other than enterotoxigenic *E. coli,* whereas the *rfbE* primer/probe set is specific for the gene encoding the O-antigen of *E. coli* O157:H7. Consequently, the stx2 primers amplify DNA from a broader range of bacteria than the *rfbE* primers. However, the one sample that was positive for the *rfbE*-based molecular beacon assay was negative with the *rfbB* conventional PCR assay. This difference may have been due to sequence heterogeneities within the *rfbB* gene of the particular *E. coli* O157:H7 present in the sample, leading to a primer mismatch with the *rfbB*-specific primers. Alternatively, the rfbB-based assay may be less sensitive than the *rfbE*-based molecular beacon assay. Nevertheless, a positive test result with both the *stx2* conventional PCR and the *rfbE* molecular beacon assays demonstrated that at least one of the environmental water samples (an urban stream affected by wastewater) was contaminated with *E. coli* O157:H7.

2.7.5 Development of Molecular Beacon-Based Real-Time PCR Assay for Detection of *Salmonella* spp.

Thermal cycling conditions using the Perkin-Elmer ABI Prism 7700 sequence Detection System for the *Salmonella* spp.-specific molecular beacon assay were: initial melting at 98°C for 10 minutes, followed by 40 cycles of melting at 95°C for 60 seconds, annealing at 60°C for 60 seconds and extension at 72°C for 60 seconds. A final extension step at 72°C for 5 minutes was included. Fluorescence measurements were recorded during each annealing step. As described above for the *E. coli* O157:H7 assay, data was automatically analyzed at the end of each PCR run and amplification plots were obtained. Amplification reactions contained 5 µL of DNA template, 45 µL of PCR master mix (5 µL of 1 X PCR buffer II [Perkin-Elmer], 3.5 mM $MgCl_2$, 500 nM of each primer [SHIMAL and SHIMAR], 200 nM dNTP mix, 0.5 U of AmpliTaq DNA Polymerase, 0.5 µM of molecular beacon probe [BHMA1], and 25 µL of water). The PCR buffer contained the fluorophore ROX (60 nmol final concentration) as the reference for normalization of the reactions. Any possible fluctuations in ROX signal are used to correct the sample signal. Ten-fold serial dilutions of DNA extracted from *Salmonella typhimurium* were used to assess the sensitivity of the assay and to generate a standard curve (Figure 2-9). A no-template control, in which sterile buffer was substituted for template DNA, was used in each experiment. The fluorescence produced by this control was subtracted from all sample signals. A plot of normalized fluorescence signal as a function of the number of PCR cycles clearly demonstrated the progression of the amplification reaction in each sample (Figure 2-9). As few as 2 CFU were detected by the *Salmonella* spp.-specific molecular beacon assay, and the C_t values decreased linearly with increasing target quantity up to 1.8×10^4 CFU (Figure 2-9).

Table 2-13. Detection of Enterotoxigenic Bacteria in Environmental Water Samples[a].

Sample type	Detection using the following PCR-based assays:		
	rfbE molecular beacon[(c)]	*stx2* conventional PCR[(d)]	*rfbB* conventional PCR[(e)]
Untreated surface water	-	+	-
Untreated surface water	-	-	-
Urban stream[(b)]	+	+	-
Urban stream[(b)]	-	+	-
Urban stream[(b)]	-	+	-
Urban stream[(b)]	-	+	-
Urban stream[(b)]	-	+	-
Positive spike control	+	+	+
Positive spike control	+	+	+
Negative control	-	-	-

[(a)] 10-L samples were processed using the GF/F filtration-enrichment method followed by DNA extraction.
[(b)] Urban stream influenced by wastewater.
[(c)] Assay run in duplicate with complete agreement between replicates.
[(d)] Results based on a combination of PCR and probing.
[(e)] Results based on PCR alone.

Five individual quantitative PCR assays were performed using the *Salmonella* spp.-specific molecular beacon assay to assess the level of intra-assay variability. Initial target concentrations (based on 10-fold serial dilutions of DNA extracted from a stock suspension of cells) ranged from 1.8 to 18,000 CFU per reaction, and each amplification was performed in duplicate. Comparison of the C_t values from the five replicate assays demonstrated that the assay was highly reproducible with little variability (Figure 2-9). Also, the rate of fluorescence increase at each target concentration was similar among the five different replicate assays. The average coefficient of variation in the calculated C_t values among the five replicate analyses for all initial target concentrations was 1.4%.

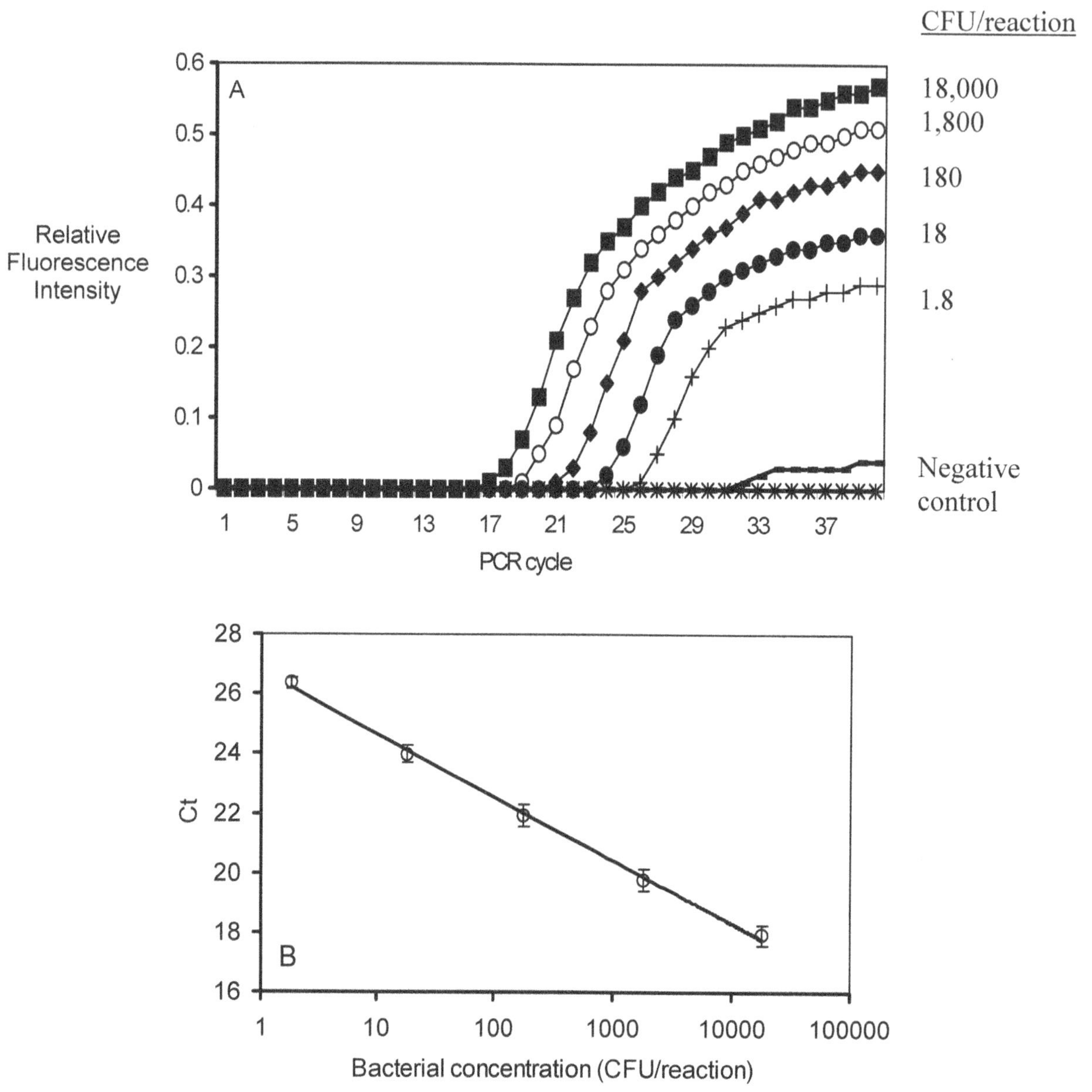

Figure 2-9. Sensitivity (A) and Standard Curve (B) of *Salmonella* spp.-specific Molecular Beacon Assay Based on Serial 10-fold Dilutions of DNA Template.
Error bars represent the standard deviation of five replicate analyses.

2.7.6 Specificity of Molecular Beacon Assay for *Salmonella* spp.

Seven different *Salmonella* species isolated from the CSDLAC culture collection were analyzed using the molecular beacon assay. The isolates were identified as *S. thompson*, *S. enteriditis*, *S. newport*, *S. heidelberg*, *S. cholerasius*, *Salmonella* sp. G1F7, and *Salmonella* sp. OC1, and all seven targets produced a strong fluorescent signal ($\Delta Rn > 0.5$ after 40 cycles). The assay was negative with *Citrobacter freundii* and *Enterobacter cloacae*.

In order to demonstrate the specificity of the molecular beacon assay for *Salmonella* spp., primers were designed to specifically amplify the corresponding region of the *himA* gene in *E. coli*. PCR products were examined through gel electrophoresis. A 122-bp product was amplified from both *E. coli* and *Salmonella* spp. templates. However, there were two base pair mismatches between the BHMA1 probe and the expected *E. coli* amplicons, so the molecular beacon probes formed unstable hybrids that dissociated during primer annealing, the stage at which fluorescence is recorded. Consequently, the molecular beacon assay did not detect *E. coli*, even though amplified product was present (Figure 2-10), thus demonstrating the ability of molecular beacons to discriminate between very similar sequences.

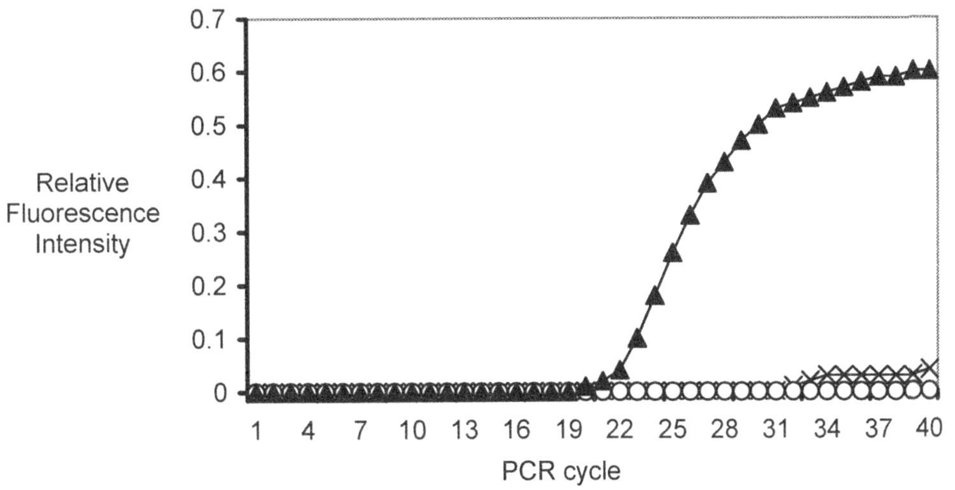

Figure 2-10. Real-Time PCR Amplification of *himA* Gene from *Salmonella* spp. (▲) and *E. coli* (○) Using Primers SHIMAL/R and EHIMAF/R, Respectively, and Probe BHMA1. ×.
No template control.

CHAPTER 3.0

PHASE 2 MONITORING PROGRAM

3.1 General Description and Goals

The second phase of the project was planned to be a field validation of the selected methods and to develop a database to relate concentration of selected indicator organisms to the probability of pathogens being present, and concurrently evaluate direct detection of a list of selected pathogens using molecular techniques. The sampling program was designed to develop a database large enough to establish sound relationships between the occurrence of pathogens detected by PCR, and culture based analyses, versus the concentrations of proposed alternative indicator organisms. The indicator organisms analyzed during the monitoring program are listed in Table 3-1. Indicator organisms were analyzed using conventional microbiological methods as described in Section 3.3.

Table 3.1. Indicator Organisms Tested During Monitoring Program.

Indicator Organisms
Total coliform
Fecal coliform
E. coli
Fecal streptococci
Enterococci
Clostridium perfringens
Somatic coliphage (CN13 host)
Male-specific coliphage (F-amp host)

Samples for bacterial pathogens were concentrated and analyzed by a combination of the methods evaluated and described in the Phase 1 portion of the project. Table 3-2 lists the bacterial pathogens included and summarizes which pathogens were analyzed by each respective method. Method specifics are described in Section 3.3.

3.2 Monitoring Sites and Sampling Program

A total of 200 samples were collected and analyzed during the monitoring phase of the study. Sample sites were selected to represent a range of anticipated water qualities, from pristine to very poor, and included rivers, lakes, and both secondary and tertiary wastewater effluents. Sample sites were also selected to represent regions of the country with different climatic conditions and water quality issues, such as combined sewer overflows. The majority of the sample sites were located in southern California, due to the logistics of collecting large volume samples. Some of these local sample locations, however, represented water from geographically large western U.S. watershed areas. Table 3-3 gives a generic breakdown of the distribution of samples collected for the study. More detailed descriptions of each sample location are provided in Appendix C.

Table 3-2. Analytical Strategies for Detecting Potential Pathogens during Phase 2 Monitoring Program.

Organism	1-L Sample Volume	10-L GF/F-DE Concentrate	
	Membrane Dissolution - PCR	Enrichment - PCR	Culture Methods
Aeromonas spp.	X	X	X[(a)]
Campylobacter jejuni	X		
Clostridium perfringens	X		X[(b)]
Helicobacter pylori	X		
Mycobacterium avium	X		
Mycobacterium intracellulare	X		
Salmonella spp.	X	X	X
Shiga toxin producing *E. coli* (STEC)	X	X	X
Shigella spp.	X	X	
Yersinia enterocolitica	X	X	

(a) *Aeromonas* was analyzed by culture methods, however, concentrations were high enough that sample concentration was not required.
(b) Included with indicator organisms, concentration not required.

Table 3-3. Percent Distribution of Samples by Generic Characterization.

Sample Characterization	Number of Samples
Surface Water:	184
Colorado River @ Water Treatment Plant Inlet	36
Calif. State Project Water @ Water Treatment Plant Inlet	36
Out of Area – Western U.S. Watershed	24
Other Regions of U.S.	16
Urban Influenced Streams – Wastewater Effluent Impacted	48
Urban Influenced Streams – No Wastewater Effluent	24
Wastewater Effluents:	16
Tertiary Chlorinated – Non Nitrified Effluent	4
Tertiary Chlorinated – Nitrified Effluent	4
Small Residential Package Plant – Secondary Chlorinated	4
Large Secondary Treatment Plant – Chlorinated Effluent	4

A simple coding system was established to identify and categorize the selected sampling locations described in Appendix C. The first three letters of the sample ID, followed by a dash, designate a specific, identifiable water category based on source. "CTE" and "CSE" indicate chlorinated filtered tertiary wastewater effluent and chlorinated secondary effluent respectively. "USW" indicates untreated surface water. This category included rivers, lakes, reservoirs and aqueducts, and generally represented samples ranging from very good to reasonably good water quality. "UIS" indicates urban influenced streams, either with or without wastewater effluent discharges. These were locations where the stream flow was significantly impacted by urban runoff and the associated by-products of densely populated urban environments. Water quality was generally poor at these locations. Letter designations following the dash designate the specific sampling locations within each general category, as listed in Appendix C.

Samples were collected following standard aseptic microbiological sampling protocols to fill sterile Nalgene 1-L bottles and 10-L carboys. Samples were then transported to the laboratory for concentration and analysis. Local samples were processed on the same day the samples were collected. Samples shipped from other locations were transported in coolers with "freeze packs". These samples were shipped via overnight express and processed on the next day.

3.3 Phase 2 Methods

3.3.1 Culture-based Methods

3.3.1.1 Conventional Indicator Organism Analyses

Total coliform (9221B), fecal coliform (9221E), *E. coli* (9221F), fecal strep (9230B) and enterococci (9230B) analyses were performed using multiple dilution tube analyses as described in the cited sections of *Standard Methods*, 20th ed.

3.3.1.2 *Clostridium perfringens*

A membrane filtration method was used for the enumeration of *Clostridium perfringens* in surface waters (Bisson and Cabelli., 1978; Armon and Payment, 1987). M-CP agar (Acumedia™) was prepared by dissolving 7.11 g m-CP agar in 89.0 mL DI H_2O. The medium was autoclaved for 15 minutes at 121°C, and cooled to 50°C. Then 0.04 g D-cycloserine, 1.0 mL of filter sterilized Polymyxin B sulfate solution (stock sol. 2.5 mg/mL), 0.2 mL of a 4.5% filter sterilized solution of $FeCl_3$-$6H_2O$, 2.0 mL of a filter sterilized 0.5% phenolphthalein diphosphate solution, and 8.0 mL of filter sterilized 0.075% Indoxyl D- Glucoside solution were aseptically added, gently mixed and 4-5 mL was dispensed into 60x15 mm sterile disposable Petri dishes. Gelatin agar, 3%, was prepared by dissolving: 1.5 g Tryptone, 1.0 g yeast extract, 3 g gelatin and 1.5 g agar in 100 mL DI H_2O, autoclaved at 121°C for 15 minutes, cooled to 50°C and 4-5 mL was dispensed into 60x15 mm sterile disposable Petri dishes. Frazier's Reagent was prepared by adding 5 g $HgCl_2$ and 20 mL concentrated HCl in 100 mL DI H_2O.

Samples were filtered through Millipore HA filters (0.45 um pore size). Sample volumes to be filtered were determined by the water quality and estimated *Clostidium perfringens* density of the water tested. Appropriate dilutions were prepared in phosphate buffered dilution water (*Standard Methods*). Filters were placed on the m-CP agar and the Petri dishes were placed in a Gas Pak pouch, which was sealed and activated per manufacturer's instructions. The pouch was submerged in a water bath at 45°C and incubated 18-24 hours. Filters with yellow colonies were presumptive positives for *Clostridium perfringens*. Filters with a countable range of 20-80 colonies were confirmed as follows: 1) Acid phosphatase test: the filter plate was passed over an open vessel containing concentrated ammonium hydroxide. Colonies of *Cl. perfringens* turn a dark pink. 2) In situ gelatin test: Test 10% of phosphatase positive filters for gelatinase activity by transferring the filter with sterile forceps from the m-CP medium to a plate of the gelatin medium. After 1-2 hours incubation at 37°C remove the filter and flood the plate with 0.1-0.5 mL of Frazier's reagent. Colonies on the filter corresponding to hydrolysis (zone of clearing) in the medium are positive. Results were reported as CFU *Cl. perfringens* per 100 mL.

3.3.1.3 *Salmonella* spp.

Two isolation procedures were used to detect *Salmonella* spp.

Salmonella: SBG Method. 10-L samples of surface water were concentrated using the 142 mm GF/F-DE method. The filter with DE was removed from the filter holder and placed aseptically into Selenite Brilliant Green broth (24.2 g/L SBG; Difco). The SBG broth was heated to 70°C in a water bath instead of boiling for 10 minutes as indicated by the manufacture's instructions. The SBG and filter were incubated at 35°C ± 0.5°C for 24 hours and then streaked onto plates of xylose-lysine desoxycholate agar (XLD; Difco) and modified lysine iron agar (m-LIA) (Rappold and Bolderdijk, 1979) for primary isolation. After streaking, the enrichment culture was returned to the incubator for another 18-24 hours. Modified lysine iron agar consisted of 34.5 g LIA (Difco), 10 g sucrose, 10 g lactose and 1.5 g bile salts No.3 (DIFCO) and 3.7 mL of 2% novobiocin (added after boiling and cooling) per liter DI water. Plates were incubated for 18-24 hours at 37°C ± 0.5°C. If the XLD and m-LIA plates were negative after 24 hr, they were restreaked from the 48-hour enrichment culture. In addition, 1mL from the corresponding 48 hr enrichment culture was inoculated into a tube of fresh SBG enrichment broth and incubated 24 hours at 37°C ± 0.5°C. The secondary enrichment tubes were streaked to XLD and m-LIA if the plates from the 48-hour enrichment cultures were negative. Salmonellae isolates were confirmed by screening typical isolates for the correct TSI, LIA and urease reaction before serological confirmation with polyvalent O antisera.

Salmonella: MSRV Method. 10-L samples of surface water were concentrated using the 142 mm GF/F-DE method. The filter with DE was removed from the filter holder and placed aseptically into Tryticase Soy broth (Difco). The TSB was incubated 24 hours at 37°C ± 0.5°C. The same nonselective TSB enrichment culture was also used to extract DNA for PCR analysis. TSB cultures were spotted onto Rappaport-Vassiliadis agar medium (MSRV), semisolid modification. MSRV was prepared by adding 31.6 g of MSRV (Difco) to 1 L deionized water and mixed well. The medium was boiled, but not autoclaved, cooled to 50°C and 1.0 mL of a 2% stock solution of Novobiocin added per liter of medium. Thick plates with approximately 25 mL each were poured. TSB was spotted onto MSRV plates by applying six, 30µL drops onto a MSRV plate. The drops were allowed to absorb into the agar for one hour at room temperature, and then incubated, inoculated side up, at 42±0.5°C for 16-18 hours in a humidified incubator. Plates were examined for the appearance of motility surrounding the spots, as evidenced by a "whitish halo" of growth. Material from the outer edge of the halo was picked with a sterile inoculating needle and streaked onto XLD agar for isolation. XLD plates were incubated for 18-24 hours at 35°±0.5C. Typical *Salmonella* spp. colonies were confirmed as described above.

3.3.1.4 *E.coli* O157:H7

10-L samples of surface water were concentrated using the 142 mm GF/F-DE method. The filter with DE was removed from the filter holder and placed aseptically into the enrichment broth, Bacto modified EC medium. Bacto modified EC medium was supplemented with 20 mg/L novobiocin and 1.12 g/L bile salts No. 3 (Difco) (Okrend et al., 1990a and 1990b). The filter and enrichment broth was incubated for 24 hours at 35°C. After incubation, the enrichment broth was streaked to BCM™ (Biosynth) selective plating medium to obtain isolated colonies. The plates were incubated at 35°C ± 0.5°C for 24 hours. *E. coli* O157:H7 produces blue-black colonies with a surrounding black precipitate. Other strains of *E.coli* produce green colonies with a brownish center. Following incubation, presumptive *E. coli* O157:H7 colonies were picked and streaked for isolation on TSA agar plates. Isolated colonies were tested for oxidase reaction, and biochemical identification was done using API 20E test strips. Selected isolates were serologically confirmed for quality assurance using *E.coli* O and H antisera. 3.3.1.5.

3.3.1.5 *Aeromonas*

A membrane filtration method was used for the enumeration of *Aeromonas hydrophila* from surface waters using Ampicillin-Dextrin Agar (ADA) as the plating medium (Havelaar *et al*., 1987). ADA was prepared by adding 5.0 g tryptone, 10 g dextrin, 2.0 g yeast extract, 3.0 g NaCl, 2.0 g KCl, 0.2 g $MgSO_4$-$7H_2O$, 0.1 g FeCl3-$6H_2O$, 0.08 g bromothymol blue, and 13.5 g agar to 1 L DI water. The medium was heated on a stir plate to dissolve agar and autoclaved at 121°C for 15 minutes. The autoclaved medium was cooled in a 45°C water bath, 0.1 g filter sterilized sodium deoxycholate and 0.01 g filter sterilized ampicillin was added per liter, and the medium was poured into 60 mm Petri dishes. Samples were filtered through Millipore HA filters (0.45 um pore size). The filters were placed on ADA and incubated aerobically at 30°C for 20-24 hours. Lemon yellow colonies 1-2 mm were picked and confirmed by oxidase reaction (+) and biochemical profile obtained using API 20E test strips.

3.3.1.6 Coliphage

Both F-Specific and somatic coliphage were tested for this study: *E. coli* F-amp (Debartolomeis and Cabelli, 1991) was used as the host for F-specific coliphage and *E. coli* CN-13 (ATCC 700609) was used as the host for somatic coliphage. Host cultures were grown overnight at 37°C in trypticase soy broth (TSB) with antibiotics added and used the day of the assay. TSB was prepared per manufacturer's instructions. Filter sterilized ampicillin and streptomycin (0.015 g/L each) was added to sterilized TSB for *E. coli* F-amp. Filter sterilized nalidixic acid (0.1 g/L) was added to TSB for *E. coli* CN-13. Log phase cultures were prepared by transferring 1 mL of an overnight culture to fresh TSB containing the appropriate antibiotic, and incubating three to four hours at 37°C. Separate analyses using a modified 5-tube MPN procedure were run for male-specific coliphage and somatic coliphage (Kott, 1966; Yanko et al. 1999). Five mL of TSB (with antibiotics for *E. coli* F-amp or CN-13 as appropriate) was dispensed into tubes for 1 mL or smaller sample volumes. 10-mL of 2X strength medium was used for 10 mL sample volumes. After the tubes of TSB were inoculated with sample, 0.1 mL of log phase host culture (F-amp for male the specific phage test or CN-13 for somatic phage test) was added to each tube. Tubes were incubated 24 hr at 37°C.

The presence/absence of phage in the MPN "enrichment" tubes was confirmed by the spot test as described by Kott (1966) and Yanko et al. (1999). After incubation, each tube was spotted onto plates of freshly prepared lawns of the appropriate host and antibiotics to confirm the presence of male-specific coliphage or somatic coliphage. Lawns were prepared by the addition of 0.5mL of log-phage host to 3 mL soft agar (30 g TSB, 7 g agar, 0.0275 g $CaCl_2$ per liter) and poured on to the respective plating media. Tryticase Soy Agar plating medium containing the appropriate antibiotics was used for bottom agar. Plates were incubated at 37°C overnight and observed for plaque formation. An MPN was computed from the resulting positive/negative tubes.

3.3.2. Molecular Methods

The molecular methods used during Phase 2 for extraction, DNA purification, PCR and probe confirmation were described in detail in Chapter 2.0, Sections 2.2-2.6.

3.4 Analysis and Discussion of Phase 2 Results

3.4.1 Monitoring Results

An extensive amount of data was developed during the monitoring phase of this project. All Phase 2 individual data are tabulated in Appendix D (Tables D-1 to D-3) and Tables 3-5 through 3-7. The indicator organism database (Table D-1) provides valuable information on the range of microbiological water qualities encountered and the interrelationships of the indicator organisms monitored, and may have value in itself, well beyond the goals of this project. Samples were collected to represent a broad range of water qualities. Individual sites were not evaluated or compared for this project.

3.4.1.1 Conventional Culture-based Analyses

Table 3.4 gives a summary of the geometric mean and range of the indicator organism concentrations for each of the basic categories of samples collected, *i.e.* chlorinated effluents, urban influenced streams (UIS) and untreated surface water (USW). Standard quality assurance practices were followed throughout the sampling program. 10% of the samples were analyzed in duplicate. Duplicate analyses are summarized in Appendix E.

Table 3-4. Geometric Mean and Range of Indicator Organism Concentrations Grouped by Sample Category.

Indicator Organism	MPN/100 mL[a] Chlorinated Effluents		Urban Streams (UIS)		Surface Waters (USW)	
	Mean	Range	Mean	Range	Mean	Range
Total coliform	(b)	<2 - 230	12,000	800 - 2,300,000	45	2 - 80,000
Fecal coliform	(b)	<2 - 50	1,100	30 - 110,000	10	2 - 17,000
E. coli	(b)	<2 - 30	810	23 - 110,000	8	2 - 17,000
Fecal streptococci	(b)	<2 - 50	1,400	50 - 180,000	7	2 - 18,000
Enterococci	(b)	<2 - 50	410	7 - 74,000	4	<2 - 1,400
Cl. perfringens	13	<0.1 - 2,900	39	5 - 950	1	<0.1 - 1,800
CN13 Phage	48	<2 – 3,000	400	23 - 13,000	5	<2 - 1,800
F-amp Phage	260	<2 – 5,000	120	2 - 16,000	3	<2 - 300

(a) *Cl. perfringens* unit is CFU/100 mL

(b) Majority of values were <2. No mean was calculated.

Viewing the indicator organism data in Appendix C, it appears that the concentrations exhibit a positive correlation, *i.e.*, all concentrations increase together, with the exception of the chlorinated effluent samples where the *Cl. perfringens* spores and coliphage were more resistant to chlorine than the other indicator organisms. The chlorinated samples were removed from the data set and the strength of the indicator organism correlations for the surface water samples was tested. First the data were tested for normality, both as reported in Appendix C, and as log transformed data (Kolmogorov-Smirnov test: SigmaStat). All indicator organism data sets failed ($p < 0.001$). A test that fails indicates that the data varies significantly from the pattern expected if the data was drawn from a population with a normal distribution. Since the data were not normally distributed, the indicator organism correlations were tested using a nonparametric test, the Spearman Rank Order Correlation. The Spearman Rank Order Correlation is used when the data are not normally distributed and/or there is not a dependent and independent variable. The Spearman correlation coefficient r_s varies between –1 and +1. A correlation coefficient near +1 indicates there is a strong positive relationship between the two variables, with both always increasing together, and a r_s value near –1 indicates there is a strong negative relationship, with one always decreasing as the other increases. A correlation coefficient of 0 indicates no relationship between the variables. The results of the Spearman Rank Order correlation test are shown in Table 3-5. All of the indicator groups tested in surface water during this study exhibited a strong positive correlation with each other.

Table 3-5. Spearman Rank Correlation Coefficient "r" Values for All Indicator Organism Pairs (n=185)[(a)].

Indicator[(b)]	T.C.	F.C.	E.C.	F.S.	Ent.	C.P.	CN13	F-amp
T.C.	1	0.89	0.88	0.89	0.86	0.80	0.80	0.77
F.C.		1	0.98	0.93	0.90	0.82	0.87	0.84
E.C.			1	0.92	0.89	0.82	0.88	0.83
F.S.				1	0.94	0.80	0.85	0.81
Ent.					1	0.79	0.82	0.79
C.P.						1	0.76	0.73
CN13							1	0.84
F-amp								1

(a) Chlorinated effluent samples not included in data.

(b) T.C.-total coliform; F.C.-fecal coliform; E.C.- *E. coli*; F.S-fecal strep; Ent.-enterococci; C.P.-*Cl. perfringens*; CN13-somatic coliphage infecting *E. coli* CN13; Famp-male specific coliphage infecting *E. coli* F-amp.

The indicator organisms were all analyzed using conventional microbiological procedures. In addition, twenty-five percent of the samples were tested for the pathogens, *Salmonella*, *E. coli* O157:H7, and *Aeromonas* using conventional culture methods. Ten-liter concentrates were tested for *Salmonella* spp. and *E. coli* O157:H7 in a presence/absence format; *Aeromonas* was analyzed quantitatively since it is a ubiquitous organism in most surface waters. The culture-based pathogen data was grouped by sample category and is shown in Tables 3-6 through 3-8.

E. coli O157:H7 was not detected in any sample from either disinfected effluent or surface water using conventional culture methods. No *Salmonella* spp. were detected in chlorinated effluents. Salmonellae were detected in every urban runoff stream sample tested and in about 35% of the untreated surface water samples collected. A few *Aeromonas* were detected in chlorinated effluents. *Aeromonas* were present in every surface water sample tested. Concentrations were notably higher in the urban runoff samples. The higher concentrations of indicators measured in the urban runoff streams (Table 3-4) was reflected by the increased numbers of pathogens detected in these samples.

Table 3-6. 10-L Samples of Disinfected Effluents Analyzed for Selected Bacterial Pathogens using Culture-based Methods.

		Positive/Negative		CFU/100mL
Sample Date	Sample Site	*Salmonella* sp.	*E.coli* O157:H7	*Aeromonas* sp.
1/8/01	CSE-J	–	–	<4
9/26/00	CSE-L	–	–	12
3/6/00	CTE-S	–	–	1
4/10/00	CTE-W	–	–	<2

Table 3-7. 10-L Samples of Urban Runoff Impacted Streams Analyzed for Selected Bacterial Pathogens using Culture-based Methods.

		Positive/Negative		CFU/100mL
Sample Date	Sample Site	*Salmonella* sp.	*E.coli* O157:H7	*Aeromonas* sp.
2/29/00	UIS-CC	+	–	1,300,000
4/24/00	UIS-CC	+	–	130,000
6/12/00	UIS-CC	+	–	140,000
8/28/00	UIS-CC	+	–	240,000
1/2/01	UIS-CC	+	–	110,000
4/4/01	UIS-CC	+	–	42,000
2/22/00	UIS-SJC	+	–	70,000
4/3/00	UIS-SJC	+	–	600,000
5/1/00	UIS-SJC	+	–	130,000
6/5/00	UIS-SJC	+	–	53,000
9/5/00	UIS-SJC	+	–	110,000
4/16/01	UIS-SJC	+	–	71,000
3/27/00	UIS-SRD	+	–	70,000
6/19/00	UIS-SRD	+	–	1,200
9/18/00	UIS-SRD	+	–	130,000
11/13/00	UIS-SRD	+	–	3,000
1/22/01	UIS-SRD	+	–	40,000
6/15/00	UIS-SRU	+	–	5,200
9/14/00	UIS-SRU	+	–	60,000
11/9/00	UIS-SRU	+	–	31,000
1/18/01	UIS-SRU	+	–	160,000

Table 3-8. 10-L Samples of Untreated Surface Water Analyzed for Selected Bacterial Pathogens using Culture-based Methods.

		Positive/Negative		CFU/100mL
Sample Date	Sample Site	*Salmonella* sp.	*E.coli* O157:H7	*Aeromonas* sp.
7/12/01	USW-B	+	–	1,600
2/14/00	USW-C	–	–	40,000
3/20/00	USW-C	–	–	1,400
5/30/00	USW-C	–	–	1,800
7/17/00	USW-C	–	–	19,000
10/2/00	USW-C	–	–	24,000
12/4/00	USW-C	+	–	1,600
2/5/01	USW-C	+	–	140
3/20/01	USW-C	+	–	1,400
6/26/00	USW-H	–	–	2,800
11/6/00	USW-M	–	–	1,600
2/27/01	USW-O	+	–	(a)
2/7/00	USW-P	–	–	7,000
3/13/00	USW-P	–	–	9,000
5/22/00	USW-P	–	–	7,000
7/10/00	USW-P	+	–	10,000
10/10/00	USW-P	–	–	50,000
12/11/00	USW-P	–	–	43,000
2/13/01	USW-P	–	–	19,000
3/13/01	USW-P	–	–	5,600
4/20/00	USW-R	–	–	80
5/11/00	USW-S	–	–	3,000
1/30/01	USW-T	+	–	800,000
7/12/01	USW-V	+	–	35,000
12/18/00	USW-W	+	–	170
10/30/00	USW-Y	–	–	4,300

(a) Lab accident.

3.4.1.2 Molecular Pathogen Data

The molecular data is tabulated in Appendix D (Tables D-2 and D-3) and summarized in Table 3-9. This table lists the pathogens analyzed and gives a summary of the total number of positive samples for each pathogen, for both the membrane dissolution method and the 10-L GF/F-DE enrichment method. The most significant result apparent from these data was that the membrane dissolution method did not have adequate sensitivity to develop the necessary pathogen occurrence data needed to correlate with indicator organism concentrations. For example, *Salmonella* were detected in only 2.8% of the samples analyzed by the membrane dissolution method, whereas it was detected in 72% of the larger volume enrichment samples analyzed by PCR. The increased detection resulting from the larger sample volumes with pre-enrichment was also apparent for the STEC and *Aeromonas*. Testing for some of the pathogens by the membrane dissolution/PCR method was discontinued after the low rate of detection was realized.

Table 3-9. Summary of All PCR Positive Samples for the Membrane Dissolution/Direct PCR Method and the 10-L Enrichment/PCR Method – 200 Samples Collected.

	Membrane Filter/Direct PCR			10-L Enrichment/PCR		
Organism	Samples Tested	Number Positive	Percent Positive	Samples Tested	Number Positive	Percent Positive
Salmonella spp.	143	4	2.8	200	144	72
Shiga toxin producing *E. coli*	57	0	0	200	68	34
Yersinia enterocolitica	57	0	0	200	3	1.5
Aeromonas hydrophila	53	29	55	200	181	90
Shigella	189	0	0	200	0	0
Clostridium perfringens	199	5	2.5	na	-	-
Campylobacter jejuni	198	3	1.5	na	-	-
Helicobacter pylori	199	0	0	na	-	-
Mycobacterium avium	199	22	11	na	-	-
Mycobacterium intracellulare	199	2	1.0	na	-	-

Footnote: (na) Not analyzed. The average volume analyzed by the MF/Direct PCR method was 0.31 mL.

Table 3-10 shows a direct comparison of the subset of samples and pathogens that were analyzed by both the membrane dissolution method and the GF/F-DE 10-L sample method. The lack of sensitivity of the membrane dissolution method is most likely primarily due to the small sample volumes actually analyzed. Initially, the method was planned to test 1-L sample volumes. However, in practice, it was rarely possible to actually filter 1-L sample volumes. The average sample volume filtered was 250 mL and ranged from 40 mL to 750 mL. Although these volumes were concentrated, only a small aliquot of the resulting concentrate was actually used in the PCR reaction, so the average equivalent volume of sample water analyzed was 0.31 mL. The volume concentrated, and equivalent volume tested for each sample is included in the data tabulations in Appendix D. If the full 1-L volumes had been concentrated for all samples, the effective sample volumes tested by PCR still would have been only 1.25 mL.

The pathogens that were detected with any notable frequency by the membrane dissolution/PCR method were those generally considered to be relatively ubiquitous in surface water, i.e. *Aeromonas* and *Mycobacterium avium* (Figure 3-1).

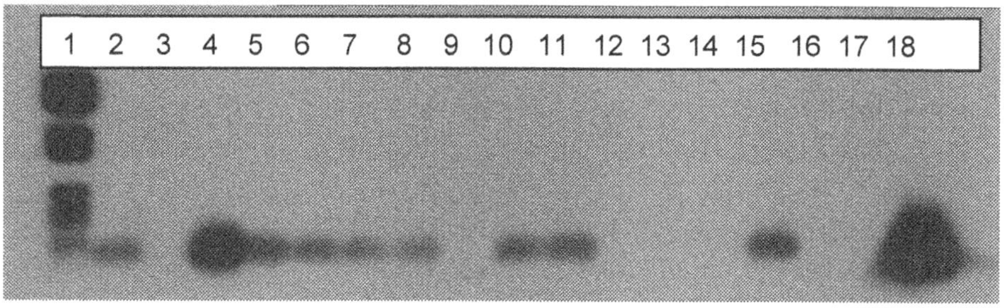

Figure 3-1. Molecular detection of *M. avium* in wastewater effluent samples using the membrane dissolution/PCR procedure (365 bp amplicon) followed by hybridization with the *M. avium*-specific probe. Lanes: 1) Molecular size standard; 2–6) secondary wastewater effluent samples; 7, 8, 10, 11, 15) tertiary treated effluents; 9, 12–14, 16) Extraction blanks; 17) PCR negative control; 18) PCR positive control.

Table 3-10. Comparison of Membrane Dissolution/PCR Method and 10-L Enrichment/PCR for the Subset of Samples Analyzed by Both Methods.

Organism	n	Number (Percent) Positive Samples MF/Direct PCR	10-L Enrichment
Salmonella spp.	144	4 (2.8%)	116 (81%)
Shiga toxin producing *E. coli*	57	0	24 (42%)
Yersinia enterocolitica	57	0	0
Aeromonas hydrophila	53	29 (55%)	50 (94%)
Shigella	190	0	0

Table 3-11 lists the subset of samples that were analyzed for *Aeromonas* by both conventional culture and by the membrane dissolution/PCR method. The first two samples in the table are chlorinated effluents and the positive PCR results most likely represent detecting dead cells. Excluding the chlorinated samples, the mean number of *Aeromonas* present in samples that tested PCR + was 1,400 CFU and the mean number of cells in the PCR negative samples was 140 CFU. These data are not robust enough to reliably estimate a detection threshold for the PCR method, and it is also possible that the molecular test and the culture test are not measuring exactly the same set of organisms. Nevertheless, it is apparent that relatively large numbers of the target pathogen must be present to reliably detect it in surface water given the small sample volumes actually analyzed.

Table 3-11. Detection of *Aeromonas* by Membrane Filter/Direct PCR Method Compared to Culture-based Concentration of *Aeromonas*.

Date	Sample	Culture Test CFU/100 mL	Membrane Filter/Direct PCR Test Vol. Analyzed - mL	CFU Analyzed	PCR Result
3/6/00	CTE-S	1	0.28	0.0028	+
4/10/00	CTE-W	<2	0.50	<0.001	+
2/29/00	UIS-CC	1,300,000	0.41	5,300	+
2/22/00	UIS-SJC	70,000	0.25	180	+
4/3/00	UIS-SJC	600,000	0.28	1,700	+
2/7/00	USW-P	7,000	0.50	35	+
3/13/00	USW-P	9,000	0.25	23	+
2/14/00	USW-C	40,000	0.63	250	–
3/20/00	USW-C	1,400	0.63	8.8	–
5/1/00	UIS-SJC	130,000	0.31	400	–
3/27/00	UIS-SRD	70,000	0.25	180	–
4/20/00	USW-R	80	0.50	0.40	–
5/11/00	USW-S	3,000	0.25	7.5	–

The inability of the membrane dissolution/PCR method to provide a usable data set on the occurrence of pathogens was a major disappointment, since one of the key components of the study plan was to correlate indicator organism densities to the probability of the Table 3-2 pathogens being detected. At the same time, the GF/F-DE concentration system with nonselective pre-enrichment followed by PCR detection proved to be a highly sensitive system for detecting the aerobic pathogens that were able to multiply in the TSB enrichment broth (Figure 3-2). In some cases the system was too sensitive, such as with the *Salmonella spp.*, where salmonellae were detected in 72% of the 10-L samples analyzed by PCR. Both the larger sample volume and the preenrichment step most likely contributed to the increased pathogen detection with the GF/F-DE method. This also raises the question of what constitutes an appropriate sample volume for an environmentally ubiquitous pathogen such as *Salmonella* spp.

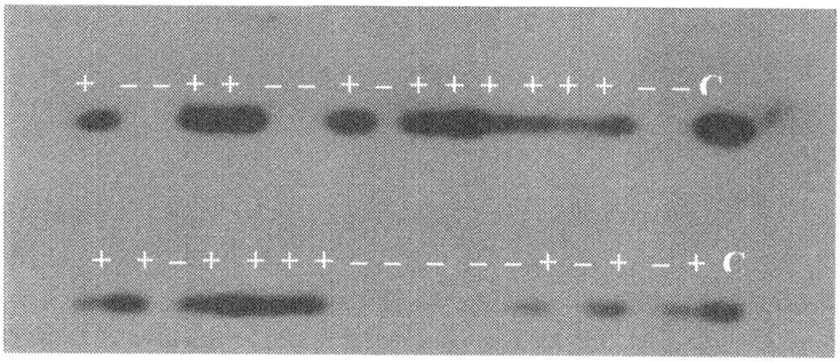

Figure 3-2. Molecular detection of *Salmonella* sp. by enrichment/PCR followed by confirmatory probe hybridization in 34 sourcewater samples. +, Positive samples (56% of these samples were positive); -, Negative samples; C, Positive control.

3.4.2 Pathogen—Indicator Organism Relationships

One primary goal of the project was to examine the relationship between indicator organism concentrations, including both standard and nonstandard indicators, and the probability of pathogenic bacteria being detected. Although the direct PCR approach proved to be inadequate to develop the necessary pathogen data, there was enough data available to explore the concept.

Two sets of pathogen data are of particular interest to examine. One is the culture-based data for *Salmonella* spp. and the other is the molecular data for STEC.

Table 3-12 summarizes the mean indicator organism concentrations for the subset of samples that were culture-positive for *Salmonella* spp. compared to the subset of samples that were culture negative for *Salmonella* spp. It is readily apparent from these data that, as a group, all indicator populations were higher in the salmonellae culture-positive samples. It is also noteworthy that individual samples were sometimes very low, or negative for indicators when *Salmonella* spp. were detected. Clearly a single sample providing a low or negative indicator count does not assure the absence of a pathogen. On the other hand, these numbers suggest that a well designed sampling program should be able to discriminate between water bodies that have a high likelihood of having salmonellae present and those with a low likelihood. A value of one-half the detection limit was used for the "less than" values in the data set to compute the means. The means for the *Salmonella*-negative samples are clearly biased by the detection limit for the indicator tests. If larger volumes had been tested for the indicators, thus lowering the detection limit, most of the geometric means reported in Table 3-12 for the *Salmonella*-negative samples would have been lower.

Table 3-12. Geometric Mean, S.D. and Range of Indicator Organisms for 10-L *Salmonella* Culture-Positive Samples Compared to 10-L *Salmonella* Culture-Negative Samples.

Organism	Unit per 100 mL	Salmonella Culture-Positive Geo. Mean & S.D.	Salmonella Culture-Positive Range	Salmonella Culture-Negative Geo. Mean & S.D.	Salmonella Culture-Negative Range
Total coliform	MPN	22,000 ± 16,000	2 - 80,000	30 ± 120	<2 - 400
Fecal coliform	MPN	340 ± 6,200	2 - 30,000	8 ± 19	<2 - 80
E. coli	MPN	270 ± 6,100	2 - 30,000	6 ± 19	<2 - 80
Fecal strep.	MPN	290 ± 5,100	<2 - 23,000	5 ± 50	<2 - 210
Enterococci	MPN	83 ± 670	<2 - 3,000	4 ± 42	<2 - 170
Cl. perfringens	CFU	20 ±100	0.4 - 440	0.5 ± 4	<0.1 - 15
CN13 Phage	MPN	190 ± 1,100	<2 - 5,000	4 ± 13	<2 - 50
F-amp Phage	MPN	59 ± 1,600	<2 - 7,000	2 ± 3	<2 - 11

Another approach to viewing these data is to look at the probability of detecting a pathogen, in this case *Salmonella* spp., at incremental concentration ranges for the indicator. Using the 10-L culture-based *Salmonella* presence/absence data, and the concentrations of *E. coli*, the data were grouped into log incremental sets, *i.e.*, 1-10, 11-100, 101-1,000, etc. The probability of detecting *Salmonella* was determined for each log-increment of the *E. coli* data. Figure 3-3 illustrates this approach. It can be seen that the probability of detecting *Salmonella* increased as the concentration of *E. coli* increased up to about 1,000 *E. coli* per 100 mL, at which point there was a 100% probability that *Salmonella* would be detected in a 10-L sample. As discussed earlier, no single sample result for the indicator provides absolute assurance about the presence or absence of the pathogen, but there is clearly increased probability of culturable *Salmonella* being present as the concentration of *E. coli* increases.

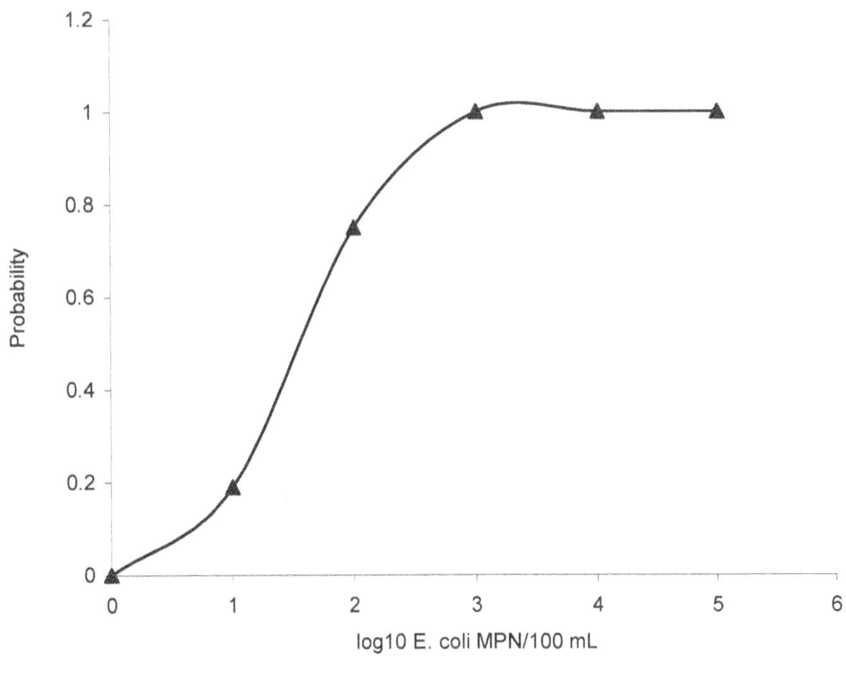

Figure 3-3. Probability of Detecting *Salmonella* by Culture in 10-L Sample versus *E. coli* Concentration per 100 mL.

Table 3-13 shows the same analysis as Table 3-12 for the samples that were positive and negative for *Salmonella* spp. using the 10-L Enrichment-PCR method. This represents a much larger data set than that in Table 3-12. Table 3-13 includes all samples except the disinfected effluents. It is interesting to note that the geometric mean concentrations for the *Salmonella*-negative samples are almost identical in the two data sets, however, the standard deviation and ranges are much greater with the PCR associated data set. The 10-L enrichment-PCR method was, in general, more sensitive than the 10-L culture analyses, however, there were 10-L samples where *Salmonella* were detected by culture and not detected by PCR, and many of those samples had high indicator organism concentrations which resulted in the larger standard deviations and ranges associated with the *Salmonella* PCR-negative indicator data set. Table 3-14 shows the relationship between the 10-L samples that were positive and negative for *Salmonella* by culture compared to PCR for the same samples. Thirty 10-L samples were *Salmonella*-positive by culture and 6 of those (20%) were negative by PCR. Conversely, 11 of 17 (65%) 10-L samples that were negative for *Salmonella* by culture, were positive by PCR.

Although the indicator organism geometric means were similar for the *Salmonella*-negative samples in both data sets, the difference between the indicator means for the *Salmonella*-positive and negative samples was not as pronounced with the PCR data set in Table 3-13, suggesting the indicators had somewhat less value for predicting the presence of salmonellae when the more sensitive molecular pathogen detection method was used. One question in comparing the data in Tables 3-13 and 3-14 would be the extent to which detecting dead salmonellae cells or DNA may have influenced the results. That remains a difficult question to address. Since the PCR data was derived from the enriched 10-L concentrates, and a small aliquot of the enrichment broth was actually analyzed, the positives most likely represent detecting cells that grew during the enrichment step.

Table 3-13. Geometric Mean, S.D. and Range of Indicator Organisms for 10-L *Salmonella* Enrichment/PCR-Positive Samples Compared to 10-L *Salmonella* Enrichment/PCR-Negative Samples.

Organism	Unit per 100 mL	*Salmonella* PCR-Positive Geo. Mean & S.D.	*Salmonella* PCR-Positive Range	*Salmonella* PCR-Negative Geo. Mean & S.D.	*Salmonella* PCR-Negative Range
Total coliform	MPN	670 ± 230,000	<2 – 2,300,000	64 ± 13,000	<2 - 80,000
Fecal coliform	MPN	110 ± 14,000	<2 - 110,000	8 ± 3,300	<2 - 17,000
E. coli	MPN	84 ± 14,000	<2 - 110,000	7 ± 3,300	<2 - 17,000
Fecal strep.	MPN	110 ± 8,300	<2 - 70,000	8 ± 5,400	<2 - 24,000
Enterococci	MPN	46± 7,100	<2 - 70,000	5 ± 3,500	<2 - 24,000
Cl. perfringens	CFU	5 ±48	<0.1 - 330	1 ± 69	<0.1 - 440
CN13 Phage	MPN	39± 890	<2 - 5,000	6 ± 1,900	<2 - 13,000
F-amp Phage	MPN	23 ± 1,900	<2 - 16,000	2 ± 52	<2 - 290

Table 3-14. Relationship between *Salmonella* Results for 10-L Culture-based Analyses and 10-L Enrichment PCR Analyses[(a)].

30 *Salmonella* Culture-Positive Samples		17 *Salmonella* Culture-Negative Samples	
No. PCR Positive	No. PCR Negative	No. PCR Positive	No. PCR Negative
24	6	11	6

(a) Chlorinated effluents not included.

A probability relationship similar to that shown in Figure 3-3 was not constructed for the enrichment-PCR *Salmonella* data because too many samples were positive (82%). With the data set skewed that far to the positive side for the pathogen, it was not possible to develop meaningful probabilities within the concentration range groupings of the indicator organism. If smaller samples had been analyzed for *Salmonella*, *i.e.* 100 mL, there most likely would have been a more even distribution of positive and negative *Salmonella* results available to develop a probability distribution. The 10-L enrichment-PCR data for the shiga-toxin producing *E. coli* (STEC), however, did distribute more evenly between positive (42%) and negative (58%) samples.

Table 3-15 summarizes the geometric means, S.D. and ranges for the indicator organisms for the STEC positive and negative samples. As seen with the *salmonellae* data, there is a difference between the indicator means for the two data sets, but again the ranges and S.D. associated with those means are great.

Table 3-15. Geometric Mean, S.D. and Range of Indicator Organisms for 10-L Shiga Toxin Producing E. coli (STEC) Enrichment/PCR-Positive Samples Compared to 10-L Salmonella Enrichment/PCR-Negative Samples.

Organism	Unit per 100 mL	STEC PCR-Positive Geo. Mean & S.D.	STEC PCR-Positive Range	STEC PCR-Negative Geo. Mean & S.D.	STEC PCR-Negative Range
Total coliform	MPN	2,800 ± 63,000	<2 - 500,000	120 ± 220,000	<2 – 2,300,000
Fecal coliform	MPN	430 ± 11,000	2 - 80,000	19 ± 11,000	<2 - 17,000
E. coli	MPN	300 ± 11,000	2 - 80,000	16 ± 11,000	<2 - 17,000
Fecal strep.	MPN	380 ± 9,500	<2 - 70,000	17 ± 17,000	<2 - 24,000
Enterococci	MPN	140 ± 8,400	<2 - 70,000	9 ± 7,400	<2 - 24,000
Cl. perfringens	CFU	14 ±78	<0.1 - 440	0.9 ± 90	<0.1 - 440
CN13 Phage	MPN	190 ± 1,900	<2 - 13,000	9 ± 370	<2 - 13,000
F-amp Phage	MPN	240 ± 1,300	<2 - 7,000	5 ± 1,500	<2 - 290

Figure 3-4(a & b) shows the probability of detecting STEC with increasing concentrations of the currently used indicators, *E. coli* and enterococci. Increasing concentrations of *E. coli* were an excellent predictor for increased probability of detecting STEC by PCR, as illustrated in Figure 3-4a. To a certain extent, that is a logical result since the STEC are a subgroup of *E. coli*. Figure 3-4b shows that increasing concentrations of enterococci also provided predictive value for the presence of STEC, although the data was not as linear as that with *E. coli*. In either case, as concentrations of the indicators increased, the probability of detecting STEC increased.

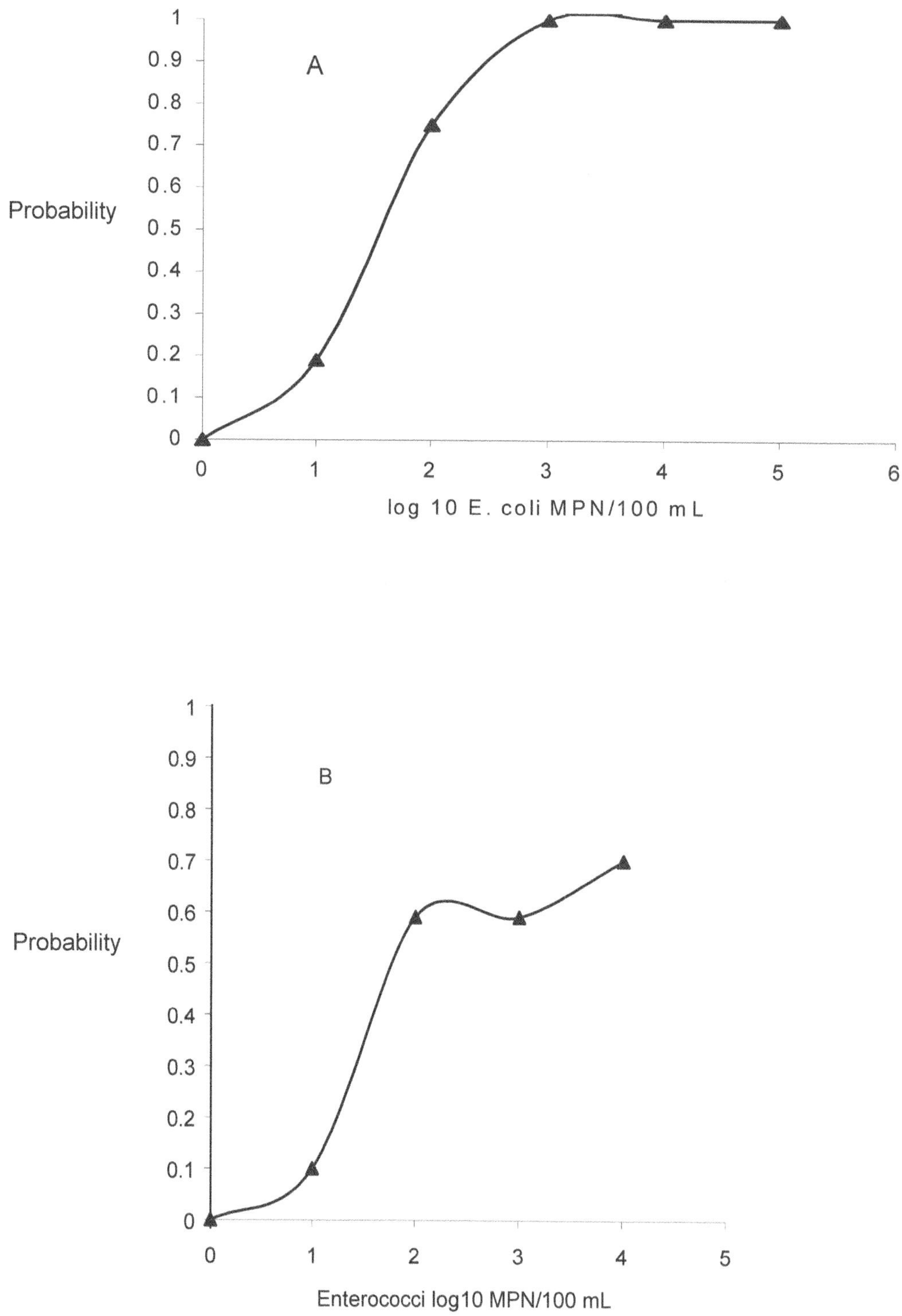

Figure 3-4. Probability of Detecting STEC by PCR in 10-L Enrichment Sample versus *E. coli* (A) and Enterococci (B) Concentrations per 100 mL.

3.5 Phase 2 Validation Testing Conclusions

The validation testing produced an extensive database on the occurrence and relative concentrations of both standard and non-standard indicator organisms in a variety of surface waters. Overall the indicators appeared to trend in unison in surface waters. All indicator concentrations were elevated in poor water quality samples and all indicator concentrations were low in good water quality samples. There was no apparent tendency for any one indicator to trend independently of the group in surface water. That was not true for chlorinated effluents, where *Clostridium perfringens* and both somatic and male-specific coliphage were more resistant to chlorination than the other indicators tested.

The 10-L GF/F concentration method proved to be a simple, low cost method to concentrate a wide range of water samples for bacterial pathogens. The resulting concentrates were suitable for culture-based analyses for *Salmonella* and *E. coli* O157:H7. In theory, the concentration method should work with any bacterial pathogen, dependent on the availability of suitable culture methods to recover the target organism from the resulting concentrate. When the GF/F concentrates were combined with an overnight non-selective enrichment step and PCR detection, the system provided a very sensitive method for detecting the aerobic pathogens. The combined system could potentially detect bacterial pathogens at very low concentrations within 24 hours. The 1-L membrane dissolution direct PCR method was not suitable for most surface water samples tested. The small equivalent sample volumes tested in most cases were inadequate to detect most of the target pathogens.

A relationship was shown between the concentration of selected indicators and 1) the probability of detecting *Salmonella* in 10-L samples by culture and 2) the probability of detecting STEC in 10-L samples by PCR. There was not enough pathogen occurrence data for other pathogens to test any other potential pathogen/indicator relationships. Indicator concentrations in single samples were not good predictors for the presence or absence of *Salmonella* or STEC, however, in conjunction with a well-designed sampling program the indicators appeared to provide useful information about the probability of the pathogens being present.

Chapter 4.0

TECHNOLOGY TRANSFER

This project originated with the premise that rapid and simple methods are needed for the collection, concentration, isolation/extraction, and identification of pathogenic bacteria in untreated surface water, finished water, and wastewater. For microbial monitoring techniques to be practical for the water and wastewater industries, they must be simple and robust enough to be performed by utility microbiology laboratories on a routine basis. Such laboratories may have only limited experience of molecular methods, if they have any experience at all, and will have neither the personnel nor resources to pursue full-scale testing and evaluation of new procedures. Therefore, one of the original objectives of this project was to demonstrate that the various sample concentration and molecular detection technologies could be transferred to previously inexperienced laboratories with a minimum of training.

Filtration of water samples through a glass fiber filter overlaid with diatomaceous earth (GF/F-DE), followed by a nonselective enrichment, DNA extraction, and specific PCR proved to be the most useful procedure for the detection of pathogens and was selected as the subject for technology transfer. The initial phase of technology transfer involved conveying protocols for the GF/F-DE filtration method from the CSDLAC research laboratory to the MWDSC research group and to a CSDLAC marine monitoring laboratory located at a different CSDLAC facility in Carson, CA. Protocols for DNA extraction and PCR/probing were transferred from MWDSC to the CSDLAC research laboratory in Whittier, CA. This involved substantial staff interchanges between the respective laboratories and hands-on training, including the transfer of QA/QC procedures.

4.1 Description of Phase 3 Technology Transfer

The technology transfer portion of the project consisted of two components, 1) a subjective evaluation of the GF/F-DE concentration method by two laboratories that had not previously used the method and 2) introduction of the molecular methods to another laboratory and a comparison of the ability of the two laboratories to produce equivalent data on split samples. The first and simpler goal was to evaluate the relative ease with which the concentration method could be introduced to laboratories that had not previously used the method and to reveal any unexpected difficulties that might be encountered. This specifically addressed the project goal of demonstrating a simple, practical, low-cost concentration method that could be utilized by basic water laboratories for concentrating bacterial pathogens. During the project, all the sample concentration and enrichment work and culture-based analyses were conducted at the CSDLAC laboratory; the molecular analyses were performed at the MWDSC laboratory. For Phase 3, the concentration method was introduced to both the MWDSC laboratory group and to another independent CSDLAC marine microbiology laboratory. The marine microbiology laboratory performs predominantly regulatory compliance testing and had no prior experience or exposure to the GF/F-DE concentration method.

Five of the sample sites tested during Phase 2 were selected for testing in Phase 3. First, five samples (10 L), one from each site, representing untreated surface water, urban-influenced streams, chlorinated secondary sewage effluent, and tertiary treated sewage effluent were processed by the GF/F-DE method combined with nonselective enrichment (18-hour incubation) and total DNA extraction by the MWDSC laboratory. Then, the same five sites were resampled, and the second set of 10-L samples was concentrated by personnel from the CSDLAC marine microbiology laboratory. These concentrates were returned to the CSDLAC research laboratory in Whittier, CA for further processing, including the 18-hour enrichment and DNA extraction. This is similar to the approach a small laboratory might use to collect and concentrate the sample and then ship the concentrate to a reference or contract laboratory for further analysis.

The second goal of Phase 3 was to assess the ability of a relatively inexperienced, new molecular laboratory to rapidly assimilate the molecular methods used in this study and produce PCR and probe results equivalent to a more experienced laboratory on a set of split samples. Each laboratory extracted DNA from the samples described above using a commercially available kit (MoBio; see Chapter 2.0) and each set of extracted DNA was then split and used by both laboratories as the template for PCR assays targeting *Salmonella* spp., *Y. enterocolitica*, *A. hydrophila*, *Shigella* spp., and Stx2-carrying enterotoxigenic coliforms (see Table 2-4 for detailed primer information). This resulted in a total of 10 samples to be analyzed for five pathogens (50 PCR analyses) at each laboratory. PCR results were confirmed by hybridization with specific probes internal to the PCR primer sets (see Table 2-7 for probe information).

4.2 Phase 3 Results

4.2.1 Large-Volume Concentration Method

Evaluation of the GF/F-DE concentration method by the MWDSC and CSDLA marine microbiology laboratories was entirely subjective. As part of the project team, the MWDSC laboratory group was conceptually familiar with the concentration method and needed minimal instruction to introduce the method into their laboratory. The MWDSC laboratory found the method easy to perform and useful for concentrating large volumes of water.

Unlike the MWDSC laboratory, the CSDLAC marine microbiology laboratory was not familiar with the GF/F-DE concentration method. Nevertheless, minimal training time was required, and the five 10-L samples were successfully concentrated the first time. No problems were encountered with use of the method. This effort reinforced the basic simplicity of the concentration method. The method should be readily adaptable in any water laboratory, and trained sampling personnel without formal microbiology training should be able to collect and concentrate samples for pathogen analysis.

4.2.2 Molecular Analyses

The two laboratories had the same results for 86.7% of samples (combining all water samples and all pathogen targets, n = 60) when results were based on PCR detection alone (Table

4-1). The level of agreement was reduced slightly to 73.3% (n = 60) based on the combined PCR/probing results (Table 4-2), with most of the differences accounted for by samples that yielded positive results at MWDSC but negative at CSDLAC. This was due to a problem obtaining detectable hybridization signal with the *A. hydrophila* probe at CSDLAC. Individual sample results for the split samples are listed in Appendix F. None of the extraction blanks processed by either laboratory during this phase of the project generated false-positive results with any of the five primer sets used, indicating the adequacy of QA/QC procedures by both the experienced and inexperienced laboratories during DNA extraction and PCR processes. The frequency of occurrence of various pathogens detected in the limited number of samples used during the technology transfer phase of the project (n = 10) demonstrated a very good correlation with the occurrence data obtained in the large field survey in which 200 samples were analyzed (Table 4-3).

Table 4-1. Comparison of PCR Results at Two Laboratories Following Technology Transfer.

		MWD	
		Positive	Negative
CSDLAC	Positive	26.7%	3.3%
	Negative	10%	60%

Table 4-2. Comparison of PCR/Probing Results Following Technology Transfer.

		MWD	
		Positive	Negative
CSDLAC	Positive	11.7%	1.7%
	Negative	25.%	61.6%

Although there was generally good concordance between the results obtained at the two laboratories following technology transfer, there were some differences (Table 4-3). These discrepancies may have resulted from: 1) the inexperienced laboratory needing a longer period of familiarization with the techniques; 2) false-positive detection by both laboratories; or 3) a heterogeneous distribution and low concentration of pathogens in the water samples such that target cells may be present in one 10-L sample but not in a second sample. False-positive detection was unlikely since all extraction blanks were negative, and dUTP and uracil-DNA glycosylase, as well as dedicated PCR facilities within the laboratory, were always employed to limit the possibility of carryover contamination from one reaction to another.

Table 4-3. Occurrence of Pathogen-Specific Sequences in Technology Transfer Samples.

	Frequency of Detection at MWD[a]		Frequency of Detection at CSDLAC[b]
Salmonella spp.	75%	[72%][c]	58%
Stx2- enterotoxigenic	25%	[34%]	33%
Yersinia enterocolitica	0	[1.5%]	0
Aeromonas hydrophila	83%	[90%]	58%
Shigella spp.	0	[0%]	0

(a) MWD results based on a combination of PCR and probing (n = 10 water samples).
(b) CSDLAC results based on PCR alone (n = 10 water samples).
(c) Numbers in parentheses represent the frequency of detection of these particular targets in the large pathogen survey (n = 200, see Chapter 3.0).

Chapter 5.0

Project Summary and Conclusions

The overall objective of this project was to develop and evaluate a method(s) that could be used to assess the bacterial pathogen load in receiving waters. A stated aim of the project was to focus on relatively simple and practical techniques that could be applied for routine monitoring purposes. The selected method would be based on either, or a combination of, 1) validating microbial indicators that predicted the presence of bacterial pathogens, or 2) direct detection of pathogens by PCR-based assays. A goal at the outset was to limit the use of defined culture-based microbiological methods, since they are cumbersome, labor-intensive, and typically demonstrate low recovery efficiencies for many pathogenic bacteria in environmental samples and are inappropriate for many fastidious microorganisms. The selected methods, combining filtration, nonselective enrichment, and PCR (with and without enrichment), were used to determine the occurrence of a suite of 10 pathogens in a large variety of environmental water samples. Conventional methods were also used to assess the occurrence of eight microbial indicators to determine those that were most closely correlated with the presence of pathogens. As part of this study, the utility of PCR-based pathogen screening methods was assessed along with its transferability to smaller water and wastewater facilities.

A simple, low-cost method involving filtration through a 142-mm diameter glass fiber-diatomaceous earth composite filter (GF/F-DE) was developed to concentrate 10-L volumes of water with turbidities up to 17 NTU. This GF/F-DE concentration system is simple to operate, requires minimal maintenance, is expected to have a long service life, and is relatively inexpensive. The single-use GF/F-DE filters represent minimal cost. The procedure was effective for recovering low numbers of bacterial pathogens from effluent and surface water samples; recovery efficiencies were 100% for samples spiked with 100 CFU, 88% for 10 CFU, and 67% for samples spiked with 1 CFU. The concentration method can be readily adapted for use by smaller utilities and would enable local personnel to collect large volume samples and reduce the volume to a size suitable for shipping to a reference or contract laboratory for pathogen analysis. Although selective culture media are available for some pathogens (*E. coli* O157:H7), the lack of available culture methods for recovering pathogens from the resulting concentrate indicated that molecular detection methods such as PCR would be most suitable for pathogen detection. Although PCR-based detection methods could be applied directly to the concentrate obtained following filtration, the methods were cumbersome and inefficient. Therefore, a nonselective enrichment was used prior to DNA extraction and PCR to enhance recovery. A primary advantage of this approach was that PCR could be used to screen for multiple pathogens in a single sample, thereby eliminating the necessity for performing multiple filtrations. Optimization of PCR primer/probe combinations, including specificity screening, for the specific detection of *Aeromonas hydrophila*, *Campylobacter jejuni*, enterotoxigenic coliforms, *Escherichia coli* O157:H7, *Helicobacter pylori*, *Mycobacterium avium* complex (*M. avium*, *M. intracellulare*, and related bacteria), *Salmonella* spp., *Shigella* spp., and *Yersinia enterocolitica* was a critical component of the methods development phase of the project.

The large-volume concentration method through GF/F-DE, combined with overnight enrichment and PCR-based detection, was demonstrated to be a sensitive method for detecting selected aerobic bacterial pathogens within 24 hours with a sensitivity of less than 10 CFU/10 L. Two hundred environmental water samples, including surface waters with and without wastewater effluent discharges and with a wide range of water quality characteristics and turbidities up to 36 NTU, were analyzed using the GF/F-DE enrichment/PCR approach: 90% were positive for *A. hydrophila*, 72% were positive for *Salmonella* spp., 34% for enterotoxigenic coliforms, and 1.5% for *Y. enterocolitica*.

An alternative PCR/probe-based approach involving direct extraction of DNA following filtration through 47-mm polycarbonate filters was also evaluated for detection of pathogens in environmental water samples. While this method had a sensitivity of 1 CFU/100 mL of effluent water, it suffered from the typical drawback of PCR inhibition by co-extracted contaminants in environmental DNA extracts. However, it was valuable for detecting some of the more fastidious pathogens that would not be detected by the enrichment approach. Out of up to 200 environmental water samples, 55% were positive for *A. hydrophila*, 12% were positive for the *M. avium* complex, with other pathogens detected at frequencies less than 3%.

As a further development of PCR-based detection technology, molecular beacon-PCR assays were developed for detection of *Salmonella* spp. and *Escherichia coli* O157:H7. This assay format combined amplification and confirmatory probing into a single homogeneous reaction and greatly reduced the analysis time compared to the conventional PCR/probe method. At least one urban stream sample was confirmed as positive for *E. coli* O157:H7 by both conventional PCR/probing and the molecular beacon-PCR assay.

A primary goal of the project was to examine the relationship between microbial indicator concentrations and the probability of detecting pathogenic bacteria in water samples. The geometric mean concentrations for the various indicators ranged from 20-22,000 per 100 mL. Positive relationships were demonstrated between the concentration of selected indicators and the probability of detecting some of the pathogens in the study. The probability of detecting *Salmonella* spp. by cultural methods clearly increased as the concentration of the indicator *E. coli* increased. Increasing concentrations of the indicator *E. coli* and enterococci were also good predictors of the presence of shiga toxin-producing coliforms (*E. coli* O157:H7 and related bacteria) based on PCR-derived detection. The results demonstrated that the approach of relating the concentration of indicator organisms to the probability of pathogens being present was a useful method to evaluate the predictive value of indicators. However, a larger database on the occurrence of other bacterial pathogens would be required to evaluate the predictive value of indicators for pathogens other than *Salmonella* and enterotoxigenic coliforms.

The voluminous research literature describing the application of a wide variety of molecular techniques to the study of environmental samples would seem to indicate that the procedures are routine and the results unquestionable. However, as demonstrated by this project, a certain amount of evaluation and optimization needs to be directed at the techniques to ensure that results are reliable and consistent, particularly with the large sample volumes necessary and low target concentrations typical of pathogens in environmental waters. A thorough understanding of the limitations of the methods is necessary to ensure that the data obtained is meaningful. Although the sample processing, concentration, and extraction procedures that precede molecular detection assays need further development before the methods can be used on a wide scale for routine monitoring purposes within the environmental field, molecular-based methods hold great promise for specific pathogen detection.

This project was divided into two primary phases: 1) methods development and evaluation, and 2) a survey of pathogen and indicator occurrence in environmental waters. The potential for technology transfer to other water laboratories was also considered. The conclusions from this investigation are summarized below:

1. Methods were developed and validated for the detection of a range of enteric bacterial pathogens in environmental waters. These methods integrated cultural and molecular techniques and can be used to screen for an array of pathogens in a single sample. Applications for these methods include pathogen monitoring programs, source water protection, and screening of effluents.

2. The presence of *A. hydrophila*, *Salmonella* spp., STEC *E. coli* (or other enterotoxigenic coliforms), *Y. enterocolitica*, and *M. avium* was confirmed in source and receiving waters.

3. The probability of detecting *Salmonella* spp. by cultural methods and shiga toxin-producing coliforms (STEC and related bacteria) by molecular methods clearly increased as the concentration of the *E. coli* and enterococci indicators increased.

4. The approach of correlating the concentration of microbial indicators to the probability of the presence of pathogens was a useful method to evaluate the predictive value of indicators.

5. The transfer of molecular-based technology from experienced to less experienced laboratories was demonstrated to be feasible. Most laboratories with trained microbiologists could be adapted to include some levels of molecular-based testing without excessive cost or extensive training.

6. The water/wastewater industries can utilize the products and results of this project to assist in the decision making process related to the introduction and implementation of molecular methods into pathogen monitoring and source water protection programs. While it is unlikely that currently used cultural methods will be replaced by molecular methods en masse, the availability of such methods increases the range of questions that can be answered regarding microbial water quality.

APPENDIX A

MICROBIAL METHODS USED DURING EVALUATION OF GF/F-DE CONCENTRATION METHOD

***Salmonella typhimurium*:** Selenite Brilliant Green broth (Difco) was used as the enrichment medium. The SBG broth was heated to 70°C in a water bath instead of boiling for 10 minutes as indicated by the manufacture's instructions. SBG was incubated at 35°C ± 0.5°C for 24 hours and then streaked onto plates of xylose-lysine desoxycholate agar (XLD; Difco) for primary isolation. After streaking, the enrichment culture was returned to the incubator for another 18-24 hours. Plates were incubated for 18-24 hours at 37°C ± 0.5°C. If the XLD plates were negative after 24 hours, they were restreaked from the 48-hour enrichment culture. In addition, 1mL from the corresponding 48-hour enrichment culture was inoculated into a tube of fresh SBG enrichment broth and incubated 24 hours at 37°C ± 0.5°C. The secondary enrichment tubes were streaked to XLD if the plates from the 48-hour enrichment cultures were negative. Salmonellae isolates were confirmed by screening typical isolates for the correct TSI, LIA and urease reaction before serological confirmation with polyvalent O antisera.

***E. coli* O157:H7**: The following method evolved during the conduct of the experiments. Bacto modified EC medium (m-EC) was used as the enrichment medium for the recovery E. coli O157:H7. Bile salts No. 3 and sodium novobiocin was added to medium achieve the selectivity of the medium. M-EC has been reported useful in the enrichment and detection of *E. coli* O157:H7 from meat, poultry products and surface waters (Okrend, et al., 1990a, Okrend, et al., 1990b, Mager, et al., 1998). The broth was incubated at 35°C for 24 hours. Following incubation, an aliquot of the enrichment broth was streaked to selective plating BCM™ medium to obtain isolated colonies. *E. coli* O157:H7 grown on BCM™ produces colonies that have a distinctive blue-black color immediately surrounded by a black precipitate. Other strains of *E. coli* produce green colonies with a brownish center. The plates were incubated at 35°C ± 0.5 for 24 hours. Following incubation, up to three target colonies were picked into 10 mL of ONPG-MUG reagent (IDEXX) and simultaneously streaked to another BCM™ plate for isolation. ONPG-MUG tubes were incubated at 35°C for 24 hours ± 2 hours. Those producing a positive ONPG reaction (yellow color) with a negative MUG reaction (no fluorescence) are considered presumptive for *E. coli* O157:H7. Identification was confirmed with an agglutination test using *E. coli* O157:H7 O and H antiserum (Difco).

***Yersinia enterocolitica*:** Samples were enriched in peptone sorbitol bile salts broth (PSB) (Weagant et al., 1983). The PSB broth was incubated up to four weeks at 4°C. PSB enrichment cultures were streaked to Yersinia Selective Agar (YSA) (Difco) and incubated 48 hours at 25°C. Presumptive colonies were screened for TSI and LAIA (lysine arginine iron agar) reactions and colony size on TSA agar after 24 hours incubation at 37°C. Final identification was confirmed by performing motility tests (48 hours) at 25°C and 37°C, and biochemical reactions after four days for ornithine decarboxylase (25°C), urease agar (25°C), and Simmon's citrate (37°C).

APPENDIX B

SELECTED ORGANISM PRIMER SPECIFICITY

Table B-1. PCR Summary for Selected Organism Primer Specificity.

Organism	Source	UNIV	SHIMA	SLTII	BHA	PL	REY	SH	UREC	MAC	PG	AERO
Target Organisms (See Table 2.4)		+	+	+	+	+	+	+	+	+	+	+
Acinetobacter calcoaceticus	ATCC #23055		-	-	-	-	-	-	-	-	-	-
Aeromonas veronii	ATCC #9071		-	-	-	-	-	-	-	-	-	-
Alcaligenes faecalis	ATCC #8750		-	-	-	-	-	-	-	-	-	-
Aquaspirillum itersonii subsp. itersonii	ATCC #12639			-	-	-	-	-	-	-	-	-
Bacillus subtilis	ATCC #6633	+	-	-	-	-	-	-	-	-	-	-
Citrobacter freundii	ATCC #8090		-	-	-	-	-	-	-	-	-	-
Clostridium beijerencke	ATCC #8260		-	-	-	-	-	-	-	-	-	-
Clostridium difficile	ATCC #9689		-	-	-	-	-	-	-	-	-	-
Comomonas testosteroni	ATCC #11996		-	-	-	-	-	-	-	-	-	-
Corynebacterium ammoniagenes	ATCC #6871		-	-	-	-	-	-	-	-	-	-
Enterobacter cloacae	ATCC #13047		-	-	-	-	-	-	-	-	-	-
Escherichia coli	ATCC #11775	+	-	-	-	-	-	-	-	-	-	-
Escherichia coli	ATCC #15597	+	-	-	-	-	-	-	-	-	-	-
Flavobacterium aquatile	ATCC #11947		-	-	-	-	-	-	-	-	-	-
Klebsiella oxytoca	ATCC #43863	+	-	-	-	-	-	-	-	-	-	-
Klebsiella pneumoniae subsp. pneumoniae	ATCC #9997	+	-	-	-	-	-	-	-	-	-	-
Klebsiella pneumoniae subsp. pneumoniae	ATCC #13883	+	-	-	-	-	-	-	-	-	-	-
Klebsiella pneumoniae subsp. pneumoniae	ATCC #27736	+	-	-	-	-	-	-	-	-	-	-
Klebsiella pneumoniae subsp.ozaenae	ATCC #11296	+	-	-	-	-	-	-	-	-	-	-
Klebsiella trevisasnii	ATCC #33558	+	-	-	-	-	-	-	-	-	-	-
Lambda bacteriophage DNA	Sigma		-				-					

Note: A positive result (+) indicates that there was sufficient DNA quantity, purity, and integrity present for subsequent PCR amplification using the species-specific primers.

Table B-1 (continued). PCR Summary for Selected Organism Primer Specificity.

Organism	Source	UNIV	SHIMA	SLTII	BHA	PL	REY	SH	UREC	MAC	PG	AERO
Target Organisms (See Table 2.4)		+	+	+	+	+	+	+	+	+	+	+
Mycobacterium kansasii	ATCC #12478	+	-	-		-	-	-	-	-	-	-
Nitrosomonas europea	ATCC #19718	+	-	-	-	-	-	-	-	-	-	-
Nitrosomonas europea	ATCC #25978	+	-	-	-	-	-	-	-	-	-	-
Phage M13	Sigma						-					
Phage PhiX-174 bacteriophage DNA	Sigma		-	-	-	-	-	-	-		-	-
Phage Q-B bacteriophage	ATCC #23631-B1						-					
Phage T7 bacteriophage DNA	Sigma	-	-	-	-	-	-	-	-		-	-
Plesiomonas shigelloides	ATCC #14029	+	-	-	-	-	-	-	-	-	-	-
Proteus vulgaris	ATCC #13315	+	-	-	-	-	-	-	-	-	-	-
Pseudomonas fluorescens	ATCC #13525	+	-	-	-	-	-	-	-	-	-	-
Pseudomonas phaesicola	ATCC #21781	+	-	-	-	-	-	-	-	-	-	-
Pseudomonas putida	ATCC #12633	+	-	-	-	-	-	-	-	-	-	-
Salmonella choleraesuis ssp. arizonae	ATCC #13314	+	+	-	-	-	-	-	-	-	-	-
Salmonella choleraesuis ssp. choleraesuis, enteritidis	ATCC #13076	+	+	-	-	-	-	-	-	-	-	-
Serratia marcescens	ATCC #13880	+	-	-	-	-	-	-	-	-	-	-
Shigella flexneri	ATCC #12022	+	-	-	-	-	-	-	-	-	-	-
Staphylococcus aureus	ATCC #25923	+	-	-	-	-	-	-	-	-	-	-
Stenotrophomonas maltophila	ATCC #13637	+	-	-	-	-	-	-	-	-	-	-
Xanthobacter autotrophicus	ATCC #43050	+	-	-	-	-	-	-	-	-	-	-
Yersinia enterocolitica	ATCC #49397	+	-	-	+	-	-	-	-	-	-	-

Note: A positive result (+) indicates that there was sufficient DNA quantity, purity, and integrity present for subsequent PCR amplification using the species-specific primers.

APPENDIX C

SAMPLE SITES

Site Code	Site Description
CSE-J	This location is a large, 350-mgd municipal wastewater treatment plant located in coastal Southern California. It provides advanced primary and partial secondary treatment of wastewaters from residential, commercial and industrial sources. The treated water from both the primary and secondary treatment processes is disinfected with chlorine. The combined flow is discharged to the Pacific Ocean through an ocean outfall system in accordance with NPDES permits. The samples are collected from chlorinated secondary effluent before it enters the ocean outfall conveyance system.
CSE-L	This water reclamation plant is a small "package" wastewater treatment plant located in the San Gabriel Mountains of Southern California. The treatment plant provides primary and secondary treatment of 100% domestic wastewater collected from an adjacent neighborhood. The secondary effluent is disinfected with chlorine. The effluent is discharged to a holding pond and used for irrigation of a neighborhood golf course.
CTE-S	This Southern California water reclamation plant provides primary, secondary and tertiary treatment for 62.5 mgd of wastewater. Effluent is currently non-nitrified. The plant serves a largely residential population of about one million. Approximately 25 mgd of the effluent is sent to percolation basins for groundwater recharge. Another 1.3 mgd is used for irrigation of two local golf courses; 1.5 mgd goes to irrigation at a landfill and nearby Memorial Park. A small amount is used by a nursery contractor. The remainder of the effluent is discharged to a river, in accordance with NPDES permits, where it eventually flows into the Pacific Ocean.
CTE-W	Site W is a Southern California water reclamation plant that provides primary, nitrified secondary and tertiary treatment for approximately 15 mgd of wastewater. The plant serves a largely residential population. On occasion, a portion of the plant discharge is diverted to a spreading basin for groundwater recharge. When not being used for recharge purposes, the effluent is discharged to a river, in accordance with NPDES permits, where it eventually flows into the Pacific Ocean.
UIS-SJC	This is a drainage channel in urban Southern California that transitions from natural creek to a mainly concrete lined channel and then to a natural bottom channel. Imported aqueduct water may be introduced into the channel for transport to groundwater recharge facilities. There is one permitted wastewater treatment plant discharging reclaimed water into the channel upstream from the

sampling point.

UIS-CC This is an urban drainage channel that originates entirely within a high density population area of Southern California. The drainage area is in excess of 17,000 acres and represents a combination of residential, business and light industrial use. The stream is predominantly a concrete lined open channel. Average dry weather flow is 10 to 20 mgd. All sources are nonpoint sources representing urban and stormwater runoff. There are no permitted discharges upstream of the sampling point for this project.

UIS-SRU This sample site is located in a diversion channel that runs adjacent to a river in the Southern California area. Sampling is done at a diversion structure leading to a 500-acre constructed wetland. Original source water (the river) is affected by both point and nonpoint sources of discharge. The river water is primarily composed of a mixture of secondary and tertiary treated effluents from a number of wastewater treatment plants upstream. In addition, there is a small groundwater component to the flow. Snowmelt from the upper portions of the watershed is minimal to non-existent. Nonpoint discharges include stormwater runoff, irrigation runoff from local urban areas and agricultural runoff from a high density of dairies located adjacent to the river and side channels. Runoff from these dairies is known to be high in nitrate, TDS, and coliform concentrations. 50% of the river is diverted into the natural bottom and vegetated diversion channel that flows approximately one mile. Midway through this channel, another wastewater treatment plant discharges an additional 1 mgd of effluent into the channel.

UIS-SRD This sampling location is a diversion channel in the Southern California area near the effluent of the 500-acre constructed wetlands. Samples are taken from a diversion structure two-thirds of the way through the constructed wetlands. Water at this point has been in the wetland system for approximately three to five days. The wetland contains native vegetation as well as numerous waterfowl and shorebirds. Water enters the wetlands from the diversion structure mentioned above. No other sources of inflow exist except for some groundwater upwelling.

USW-P Water associated with USW-P is untreated surface water transported by aqueduct from a major river-delta system in Northern California to water treatment facilities in Southern California. Following aqueduct transport, the water may be impounded in reservoirs that can be affected by point or nonpoint discharge; the water is subsequently transported to the treatment facilities by pipeline. The sample location is at the influent of a drinking water treatment plant receiving USW-P water transported by pipeline from reservoirs. The samples are collected from a sample tap providing 100% USW-P to pilot- and demonstration-scale treatment systems.

USW-R This sample location is the effluent pipeline of a deep impoundment drinking water reservoir that is also used for recreation. This reservoir is at the terminal end of the west branch of the USW-P aqueduct system. Both point and nonpoint pollution sources affect the reservoir. Point sources include floating public toilets that accidentally discharge into the reservoir intermittently. Nonpoint sources are dispersed and include runoff from agricultural areas, swim beaches, boating, forest roads, and public campgrounds. The samples are collected as grab samples immediately prior to pipeline distribution to water treatment plants.

USW-S USW-S is the effluent of a shallow impoundment drinking water reservoir. This reservoir is located at the terminal end of the east branch of the USW-P aqueduct system. Both point and nonpoint pollution sources affect the reservoir. Point sources include a wastewater treatment plant that discharges into the reservoir under NPDES permit. Nonpoint sources are dispersed and include run-off from urban areas, swim beaches, boating, roadways, forest roads, and public campgrounds. The samples are collected as grab samples immediately prior to pipeline distribution to water treatment plants.

USW-B This site is located at the headwaters of the USW-P aqueduct system, approximately one downstream from the pumps providing water to the aqueduct. The samples are collected as grab samples at mid-channel from a bridge. The aqueduct may be affected by nonpoint sources during transport.

USW-V This is a river feeding into the system providing water to the USW-P aqueduct system. The river flow at this point is heavily influenced by agricultural practices and irrigation discharges that dominate the course of the river and its tributaries from their points of discharge from reservoirs located upstream. The samples were collected as grab samples approximately eight feet into the stream flow from the riverbank.

USW-C Water associated with USW-C is untreated drinking water transported by aqueduct from a major river system in Southern California to water treatment facilities in Southern California. The source water may be affected by both point and nonpoint discharge prior to entry into the aqueduct system; following aqueduct transport, the water is impounded in reservoirs that can be affected by nonpoint discharge, and subsequently transported to the treatment facilities by pipeline. The sampling location is the influent of a drinking water treatment plant receiving USW-C water. The samples are collected from a dedicated sample tap installed at the head of the plant, prior to mixing and treatment.

USW-H This sample location is near the terminal end of a shallow impoundment reservoir located on a major river system in Southern California. This reservoir is affected by both point and nonpoint pollution sources. Point sources include a wastewater treatment plant that discharges into the reservoir under NPDES permit. Nonpoint sources are dispersed and include runoff from urban areas, swim beaches, boating, roadways, and public campgrounds. The samples are collected as grab samples immediately prior to the intake of the USW-C aqueduct system.

USW-T This sampling location is a nontidally influenced stream located along the southern U.S. coast at the Gulf of Mexico. Both point and nonpoint sources of pollution affect the river. Point sources include 57 permitted discharges (48 domestic and nine industrial for a total of 58.23 mgd). Nonpoint sources are roadways and urban stormwater runoff. Current monitoring of this area indicates elevated fecal coliform bacteria levels that exceed criteria for contact recreation use of this water. In addition, elevated levels of nitrite plus nitrate nitrogen, total phosphorus, and orthophosphorus occur here. Because the level of dissolved silver exceeds the acute criterion, the water has been categorized as providing "partial support of aquatic life use." Samples were collected from a bridge at a busy roadway in a large populated/industrial city in the bay system approximately 36 miles upstream of the coastline.

USW-W This sample location is on a river located in the Pacific Northwestern U.S. Headwaters for this river originate in the Cascade Mountain range. The drainage area is designated as an "Uncontrolled Watershed," and its total area covers approximately 3,093 square miles or 1.98 million acres. The river is affected by both point and nonpoint sources of pollution. Point sources include three municipal secondary wastewater plants that discharge into the river under NPDES permits. Nonpoint sources include widely dispersed runoff from roadways, urban areas, forest roads, agricultural activities, and private septic tanks. Samples are collected from a raw water tap located in the water treatment plant. The nearest municipal secondary wastewater plant discharge is located approximately two miles upstream. The raw water fecal coliform levels range from a low of <2 to a high of 170. Turbidity ranges from lows of 2 NTU to a high of 640 NTU. There are seasonal variations in water quality. Flood season occurs October through March; during this period turbidity, colloidal color, and river levels can change dramatically in less than 12 hours.

USW-Y This sampling location is on a stream located in the northeast U.S. The watershed size is about 100 acres. The stream is affected by nonpoint sources of pollution, with no point sources identified within its watershed. The watershed is within a suburban setting, with medium density residential and commercial office park land uses. There are no agricultural activities within the watershed. Samples were collected approximately 100 yards upstream from a receiving body of water, at a 90° V-notch control structure.

USW-O USW-O is a river located in the Midwestern U.S. It is affected by both point and nonpoint sources of pollution. Point sources include four major wastewater treatment plants and numerous Combined Sewer Overflows (CSOs) that discharge to the river under NPDES permits. CSOs are generally only active during rain events and discharge diluted raw sewage into the river. Nonpoint sources include stormwater discharge from urban runoff and agricultural activities. Water quality data suggest that pathogens, including bacteria, are likely the single most frequent parameter in violation of Commission water quality Standards. Samples for this project were collected in the downtown area of the city from the riverbank.

USW-M This location is the effluent of a drinking water reservoir. This reservoir is at the terminal end of the USW-C aqueduct system prior to treatment. There is no point source impact at this reservoir, but it may be affected by nonpoint sources such as runoff.

APPENDIX D

PHASE 2 DATA TABLES

Table D-1. Phase 2 Indicator Data Grouped by Sampling Sites.

Sample Date	Sample Site	Total coliform MPN/100mL	Fecal coliform MPN/100mL	*E.coli* MPN/100mL	Fecal strep MPN/100mL	Enterococcus MPN/100mL	*Clostridium perfringens* CFU/100ml	CN13[a] coliphage MPN/100mL	F-amp[b] coliphage MPN/100mL
6/27/00	CSE-J	230	50	30	2	<2	2,900	3,000	500
9/7/00	CSE-J	2	2	<2	2	<2	150	500	5,000
1/8/01	CSE-J	2	<2	<2	<2	<2	680	800	800
7/18/01	CSE-J	<2	<2	<2	<2	<2	530	230	700
4/25/00	CSE-L	230	13	13	50	50	180	<2	<2
6/6/00	CSE-L	80	<2	<2	2	2	4	<2	<2
9/26/00	CSE-L	2	<2	<2	<2	<2	1	<2	<2
2/6/01[c]	CSE-L	6	<2	<2	<2	<2	9	<2	<2
3/6/00	CTE-S	<2	<2	<2	<2	<2	1.5	30	nd
3/22/00	CTE-S	<2	<2	<2	<2	<2	4.7	4	350
7/11/00	CTE-S	2	<2	<2	<2	<2	2.3	2	80
11/2/00	CTE-S	<2	<2	<2	<2	<2	0.3	<2	23
3/21/00	CTE-W	<2	<2	<2	<2	<2	6	2	300
4/10/00	CTE-W	8	<2	<2	<2	<2	0.5	<2	30
9/6/00	CTE-W	2	<2	<2	<2	<2	<0.1	<2	230
2/15/01[c]	CTE-W	2	<2	<2	<2	<2	5	11	170
2/9/00	UIS-CC	2,300	30	23	300	230	75	240	130
2/28/00	UIS-CC	110,000	22,000	22,000	17,000	5,000	160	2,400	500
2/29/00	UIS-CC	7,000	800	800	1,700	44	150	300	230
3/1/00	UIS-CC	8,000	1,700	1,700	3,000	110	120	70	280
3/30/00	UIS-CC	500,000	23,000	23,000	17,000	3,000	50	800	500
4/12/00	UIS-CC	800	110	110	220	220	23	110	230

Table D-1. Phase 2 Indicator Data Grouped by Sampling Sites, cont'd.

Sample Date	Sample Site	Total coliform MPN/100mL	Fecal coliform MPN/100mL	*E.coli* MPN/100mL	Fecal strep MPN/100mL	Enterococcus MPN/100mL	*Clostridium perfringens* CFU/100ml	CN13[a] coliphage MPN/100mL	F-amp[b] coliphage MPN/100mL
4/24/00	UIS-CC	13,000	8,000	2,800	300	300	330	1,300	130
5/25/00	UIS-CC	2,300,000	110,000	110,000	30,000	17,000	100	2,400	16,000
6/2/00	UIS-CC	1,100	800	800	1,300	220	43	500	470
6/12/00	UIS-CC	30,000	3,000	3,000	2,300	500	30	500	220
7/12/00	UIS-CC	17,000	17,000	1,100	5,000	5,000	130	300	300
8/28/00	UIS-CC	5,000	800	300	1,100	170	76	230	300
10/3/00	UIS-CC	7,000	1,300	1,300	500	220	60	300	70
10/16/00	UIS-CC	23,000	1,300	1,300	3,000	280	75	300	27
11/1/00	UIS-CC	23,000	1,700	1,700	1,200	600	68	1,700	70
12/5/00	UIS-CC	8,000	2,300	1,300	500	300	67	500	800
1/2/01	UIS-CC	17,000	1,300	1,300	500	300	280	1,100	3,000
1/23/01	UIS-CC	30,000	2,300	2,300	2,300	2,300	112	500	14
2/7/01[c]	UIS-CC	30,000	15,000	15,000	24,000	24,000	77	220	100
3/14/01	UIS-CC	8,000	1,300	1,300	1,300	280	47	230	30
3/19/01	UIS-CC	3,000	170	170	110	110	38	1,300	23
4/4/01[c]	UIS-CC	15,000	1250	1250	220	190	76	800	84
5/8/01[c]	UIS-CC	9500	500	500	600	220	30	320	120
5/14/01	UIS-CC	11,000	700	700	500	230	45	300	80
2/15/00	UIS-SJC	30,000	2,300	2,300	23,000	5,000	170	1,300	20
2/22/00	UIS-SJC	30,000	30,000	30,000	23,000	3,000	90	3,000	5,000
2/25/00	UIS-SJC	130,000	8,000	8,000	8,000	2,200	130	5,000	3,000
3/14/00	UIS-SJC	2,200	900	700	300	170	22	700	50
3/29/00	UIS-SJC	9,000	500	500	140	140	12	23	80
4/3/00	UIS-SJC	8,000	130	130	230	230	16	30	2
4/19/00	UIS-SJC	500,000	80,000	80,000	70,000	70,000	140	5,000	5,000
5/1/00	UIS-SJC	2,300	300	300	500	300	21	300	500

Table D-1. Phase 2 Indicator Data Grouped by Sampling Sites, cont'd.

Sample Date	Sample Site	Total coliform MPN/100mL	Fecal coliform MPN/100mL	*E.coli* MPN/100mL	Fecal strep MPN/100mL	Enterococcus MPN/100mL	*Clostridium perfringens* CFU/100ml	CN13[a] coliphage MPN/100mL	F-amp[b] coliphage MPN/100mL
5/24/00	UIS-SJC	2,800	500	70	800	170	10	230	500
6/5/00	UIS-SJC	1,700	220	90	230	50	13	50	17
6/26/00	UIS-SJC	5,000	800	70	170	500	26	50	80
7/27/00	UIS-SJC	14,000	500	500	1,700	700	14	300	170
8/31/00	UIS-SJC	6,000	2,300	1,300	5,000	45	23	500	3,000
9/5/00	UIS-SJC	30,000	3500	500	1,700	280	80	700	500
9/20/00	UIS-SJC	13,000	130	23	5,000	3,000	6	1,100	500
10/5/00	UIS-SJC	70,000	11,000	11,000	23,000	3,000	26	1,100	110
10/17/00	UIS-SJC	30,000	1400	1,100	3,000	60	56	800	300
10/26/00	UIS-SJC	5,000	3,000	3,000	5,000	1,300	57	2,400	30
11/28/00	UIS-SJC	1,700	500	500	300	70	26	230	110
12/6/00	UIS-SJC	1,300	170	110	500	17	27	90	30
1/10/01	UIS-SJC	22,000	700	300	2,300	300	33	230	<2
2/21/01	UIS-SJC	22,000	2,300	2,300	3,000	110	70	500	500
3/06/01[c]	UIS-SJC	200,000	27,000	22,000	180,000	74,000	950	2,400	370
4/16/01[c]	UIS-SJC	7,200	470	175	750	175	19	30	130
3/27/00	UIS-SRD	2,200	800	800	130	130	38	800	130
4/17/00	UIS-SRD	13,000	80	80	230	80	32	30	4
5/15/00	UIS-SRD	13,000	220	170	300	170	16	30	17
6/19/00	UIS-SRD	3,000	170	170	700	33	5	300	17
7/24/00	UIS-SRD	30,000	110	110	50	22	28	170	80
8/21/00	UIS-SRD	17,000	500	220	1400	170	25	300	4
9/18/00	UIS-SRD	11,000	3,000	3,000	1,700	7	12	2,400	800
10/23/00	UIS-SRD	5,000	1,700	1,700	3,000	800	14	1,700	300
11/13/00	UIS-SRD	2,300	500	500	230	130	15	1,300	50

Table D-1. Phase 2 Indicator Data Grouped by Sampling Sites, cont'd.

Sample Date	Sample Site	Total coliform MPN/100mL	Fecal coliform MPN/100mL	*E.coli* MPN/100mL	Fecal strep MPN/100mL	Enterococcus MPN/100mL	*Clostridium perfringens* CFU/100ml	CN13[a] coliphage MPN/100mL	F-amp[b] coliphage MPN/100mL
12/18/00	UIS-SRD	17,000	2,200	2,200	2,300	120	26	5,000	110
1/22/01	UIS-SRD	7,000	1,100	1,100	700	500	38	5,000	80
2/12/01	UIS-SRD	17,000	2,200	2,200	2,300	2,000	40	13,000	290
3/23/00	UIS-SRU	5,000	1,100	1,100	300	300	30	80	2
4/13/00	UIS-SRU	3,000	130	130	230	230	34	50	21
5/11/00	UIS-SRU	23,000	230	230	1,300	500	23	80	80
6/15/00	UIS-SRU	13,000	170	170	1,700	1,100	43	50	130
7/20/00	UIS-SRU	17,000	2,300	800	2,300	1,300	42	<2	4
8/17/00	UIS-SRU	22,000	170	170	1,700	30	17	700	300
9/14/00	UIS-SRU	11,000	700	500	500	300	27	130	7,000
10/19/00	UIS-SRU	23,000	300	300	3,000	3,000	12	230	220
11/9/00	UIS-SRU	5,000	1,300	1,300	1,100	700	10	220	34
12/14/00	UIS-SRU	11,000	300	300	800	300	11	700	8
1/18/01	UIS-SRU	8,000	500	500	2,300	2,300	41	500	23
2/08/01[c]	UIS-SRU	6,500	290	220	1,400	1,400	14	30	14
5/9/00	USW-B	23	<2	<2	<2	<2	5	<2	30
8/7/00	USW-B	30	17	7	8	2	6	2	4
12/19/00	USW-B	300	300	300	170	110	2	220	50
7/12/01	USW-B	80	7	7	8	4	5	<2	<2
2/10/00	USW-C	30	30	30	50	22	0.7	27	80
2/14/00	USW-C	50	30	17	23	23	0.5	23	11
2/23/00	USW-C	50	50	50	50	50	0.9	50	17
2/24/00	USW-C	50	50	30	30	17	1.2	80	170
3/8/00	USW-C	30	30	30	30	30	0.3	23	17
3/9/00	USW-C	23	23	23	30	11	0.5	23	8

Table D-1. Phase 2 Indicator Data Grouped by Sampling Sites, cont'd.

Sample Date	Sample Site	Total coliform MPN/100mL	Fecal coliform MPN/100mL	*E.coli* MPN/100mL	Fecal strep MPN/100mL	Enterococcus MPN/100mL	*Clostridium perfringens* CFU/100ml	CN13[a] coliphage MPN/100mL	F-amp[b] coliphage MPN/100mL
3/20/00	USW-C	23	23	<2	30	30	0.3	17	8
3/28/00	USW-C	50	50	50	23	13	1	13	22
4/11/00	USW-C	170	170	110	110	21	0.6	50	9
4/26/00	USW-C	80	50	14	23	8	0.2	7	2
5/2/00	USW-C	110	50	17	22	22	0.5	22	2
5/3/00	USW-C	50	14	14	23	23	0.4	8	4
5/18/00	USW-C	41	27	27	4	4	0.1	<2	2
5/30/00	USW-C	130	6	6	<2	<2	0.7	2	4
6/14/00	USW-C	12	<2	<2	4	0	0.8	<2	<2
6/21/00	USW-C	4	2	2	2	<2	0.6	<2	<2
7/17/00	USW-C	14	<2	<2	2	<2	0.5	<2	<2
7/26/00	USW-C	4	<2	<2	<2	<2	0.2	<2	<2
9/28/00	USW-C	400	2	2	<2	<2	0.3	<2	<2
10/2/00	USW-C	340	<2	<2	4	2	0.2	<2	<2
10/4/00	USW-C	1,100	<2	<2	<2	<2	0.4	<2	<2
10/25/00	USW-C	700	<2	<2	<2	<2	0.1	<2	<2
11/30/00	USW-C	130	12	12	8	<2	0.3	170	<2
12/4/00	USW-C	23	23	23	30	8	0.4	50	23
1/4/01	USW-C	110	80	80	30	30	1.2	50	8
1/29/01[c]	USW-C	7	3	3	5	3	0.8	8	2
1/31/01	USW-C	50	11	11	17	8	0.9	22	4
2/5/01[c]	USW-C	10	8	8	12	3	0.8	11	3
3/20/01[c]	USW-C	230	51	51	24	11	2	150	28
4/19/01[c]	USW-C	260	15	15	<2	<2	0.6	18	5
5/9/01	USW-C	7	<2	<2	4	2	0.2	9	<2
6/5/01	USW-C	17	2	2	<2	<2	0.2	<2	<2
6/13/01	USW-C	8	4	4	<2	<2	0.1	<2	<2

Table D-1. Phase 2 Indicator Data Grouped by Sampling Sites, cont'd.

Sample Date	Sample Site	Total coliform MPN/100mL	Fecal coliform MPN/100mL	*E.coli* MPN/100mL	Fecal strep MPN/100mL	Enterococcus MPN/100mL	*Clostridium perfringens* CFU/100ml	CN13[a] coliphage MPN/100mL	F-amp[b] coliphage MPN/100mL
6/19/01	USW-C	22	<2	<2	4	<2	0.1	<2	<2
6/26/01	USW-C	26	<2	<2	<2	<2	0.5	<2	<2
7/10/01	USW-C	8	<2	<2	<2	<2	0.6	<2	<2
4/27/00	USW-H	50	2	2	7	<2	<0.1	<2	<2
6/28/00	USW-H	70	8	2	8	2	0.2	<2	<2
8/29/00	USW-H	220	27	22	30	2	0.1	<2	<2
5/30/01[c]	USW-H	75	2	2	2	2	0.2	<2	<2
11/6/00	USW-M	23	8	8	<2	<2	0.1	4	<2
1/8/01	USW-M	50	30	30	13	8	0.7	50	<2
3/27/01[c]	USW-M	60	24	14	16	16	0.4	60	20
5/14/01[c]	USW-M	12	6	5	2	<2	<0.1	19	3
4/18/00	USW-O	800	130	130	80	80	110	80	11
6/13/00	USW-O	80	11	11	8	8	23	4	<2
9/12/00	USW-O	8,000	2,200	1,400	130	130	56	300	50
2/27/01[c]	USW-O	290	40	30	32	8	73	340	13
2/7/00	USW-P	13	8	8	7	4	0.6	2	<2
2/8/00	USW-P	4	4	4	11	7	0.5	<2	<2
2/16/00	USW-P	11	2	2	2	2	0.7	<2	2
2/17/00	USW-P	11	2	<2	2	2	0.1	<2	<2
3/2/00	USW-P	23	4	4	4	4	2.8	<2	2
3/7/00	USW-P	13	13	13	4	4	0.9	<2	2
3/13/00	USW-P	14	9	9	4	2	0.3	17	<2
4/4/00	USW-P	<2	<2	<2	<2	<2	0.5	<2	<2
4/5/00	USW-P	2	<2	<2	<2	<2	0.3	<2	<2
4/6/00	USW-P	2	<2	<2	<2	<2	0.1	<2	<2

Table D-1. Phase 2 Indicator Data Grouped by Sampling Sites, cont'd.

Sample Date	Sample Site	Total coliform MPN/100mL	Fecal coliform MPN/100mL	*E.coli* MPN/100mL	Fecal strep MPN/100mL	Enterococcus MPN/100mL	*Clostridium perfringens* CFU/100ml	CN13[a] coliphage MPN/100mL	F-amp[b] coliphage MPN/100mL
5/22/00	USW-P	<2	<2	<2	<2	<2	<0.1	<2	<2
5/31/00	USW-P	2	<2	<2	<2	<2	0.1	<2	<2
6/1/00	USW-P	4	<2	<2	<2	<2	0.1	<2	<2
6/7/00	USW-P	2	<2	<2	<2	<2	0.2	<2	<2
6/8/00	USW-P	2	<2	<2	<2	<2	0.2	<2	<2
6/20/00	USW-P	8	<2	<2	<2	<2	0.5	<2	<2
6/27/00	USW-P	4	<2	<2	<2	<2	<0.1	<2	<2
7/10/00	USW-P	2	2	2	<2	<2	0.2	7	<2
7/19/00	USW-P	<2	<2	<2	<2	<2	0.2	<2	<2
8/24/00	USW-P	2	<2	<2	<2	<2	0.6	<2	<2
9/21/00	USW-P	17	<2	<2	<2	<2	0.4	<2	<2
10/10/00	USW-P	23	8	8	<2	<2	1	2	<2
10/18/00	USW-P	30	2	2	2	2	0.1	<2	<2
10/30/00	USW-P	230	<2	<2	2	2	0.7	2	<2
11/15/00	USW-P	50	22	22	2	2	2	<2	2
12/11/00	USW-P	13	8	8	4	4	3	4	2
1/16/01	USW-P	17	4	2	<2	<2	1.4	2	2
1/24/01[c]	USW-P	23	7	7	6	6	2	30	9
2/13/01[c]	USW-P	20	11	8	10	8	15	3	<2
2/26/01	USW-P	4	2	2	4	<2	10	4	<2
3/7/01	USW-P	4	2	2	<2	<2	17	<2	<2
3/13/01	USW-P	13	4	4	<2	<2	5.5	2	<2
4/5/01	USW-P	13	<2	<2	<2	<2	11	<2	<2
4/24/01[c]	USW-P	11	3	3	<2	<2	9	<2	<2
4/30/01[c]	USW-P	460	2	2	<2	<2	3	2	<2
5/7/01[c]	USW-P	<2	<2	<2	<2	<2	2	<2	<2

Table D-1. Phase 2 Indicator Data Grouped by Sampling Sites, cont'd.

Sample Date	Sample Site	Total coliform MPN/100mL	Fecal coliform MPN/100mL	*E.coli* MPN/100mL	Fecal strep MPN/100mL	Enterococcus MPN/100mL	*Clostridium perfringens* CFU/100ml	CN13[a] coliphage MPN/100mL	F-amp[b] coliphage MPN/100mL
3/16/00	USW-R	230	130	130	23	13	1.2	130	230
4/20/00	USW-R	130	30	30	50	50	0.3	50	8
7/20/00	USW-R	50	2	2	2	2	0.4	23	23
11/16/00	USW-R	80	30	30	23	23	0.5	230	8
3/15/00	USW-S	2	<2	<2	<2	<2	0.1	<2	<2
5/11/00	USW-S	23	23	13	<2	<2	<0.1	<2	<2
10/11/00	USW-S	2	<2	<2	<2	<2	0.1	<2	<2
6/20/01	USW-S	38	2	<2	<2	<2	1.2	<2	<2
5/9/00	USW-T	5,000	500	500	80	80	52	50	140
9/26/00	USW-T	40,000	1,300	1,300	3,000	140	46	130	17
1/30/01[c]	USW-T	80,000	17,000	17,000	18,000	480	440	1,800	200
5/29/01[c]	USW-T	35,000	4000	800	1,600	60	40	120	200
5/9/00	USW-V	800	500	70	110	30	30	30	300
8/7/00	USW-V	2,200	170	8	230	80	10	8	8
12/19/00	USW-V	3,000	3,000	700	17	4	6	23	170
7/12/01	USW-V	500	17	17	30	4	14	23	2
7/24/00	USW-W	130	50	22	7	4	1.5	<2	2
9/19/00	USW-W	60	30	13	30	17	2	800	2
12/18/00	USW-W	60	23	23	17	7	5	13	<2
5/21/01[c]	USW-W	46	6	6	6	2	0.8	3	<2
5/23/00	USW-Y	300	80	80	170	170	4	2	4
7/17/00	USW-Y	5,000	800	800	5,000	1400	31	17	2
10/30/00	USW-Y	400	80	80	210	170	0.8	8	<2
4/02/01[c]	USW-Y	1950	66	66	370	10	10	3	2

[a]Somatic coliphage that infect *E. coli* host CN-13.
[b]Predominantly male specific coliphage that infect *E. coli* host F-amp.
[c]Reported values for this date are the average of duplicate analyses.

Table D-2. Phase 2 Membrane Dissolution PCR Presence/Absence Data Grouped by Sampling Sites.

Sample Date	Sample Site	Sample Volume Filtered L	Effective Volume Tested mL	*Clostridium perfringens*	*C. jejuni*	*H. pylori*	*Mycobacterium* *avium*	*Mycobacterium* *intracellulare*	*Salmonella*	STEC (Stx2)	*Yersinia enterocolitica*	*Aeromonas*	*Shigella*
6/27/00	CSE-J	0.080	0.10	+	–	–	+	–	–	ø	ø	ø	–
9/7/00	CSE-J	0.070	0.09	–	–	–	+	–	+	ø	ø	ø	–
1/8/01	CSE-J	0.085	0.11	–	–	–	+	–	ø	ø	ø	ø	–
7/18/01	CSE-J	0.055	0.07	–	–	–	–	–	ø	ø	ø	ø	–
4/25/00	CSE-L	0.140	0.18	–	–	–	–	–	–	–	–	–	–
6/6/00	CSE-L	0.400	0.50	–	–	–	–	–	–	ø	ø	ø	–
9/26/00	CSE-L	0.255	0.32	–	–	–	–	–	–	ø	ø	ø	–
2/6/01	CSE-L	0.150	0.19	–	–	–	–	–	ø	ø	ø	ø	–
3/6/00	CTE-S	0.220	0.28	+	–	–	+	–	–	–	–	+	–
3/22/00	CTE-S	0.250	0.31	–	–	–	+	–	–	–	–	+	–
7/11/00	CTE-S	0.225	0.28	–	–	–	+	–	+	ø	ø	ø	–
11/2/00	CTE-S	0.255	0.32	–	–	–		–	–	ø	ø	ø	–
3/21/00	CTE-W	0.340	0.43	–	–	–	+	–	–	–	–	+	–
4/10/00	CTE-W	0.400	0.50	–	–	–	+	–	+	–	–	+	–
9/6/00	CTE-W	0.250	0.31	–	–	–	+	–	–	ø	ø	ø	–
2/15/01	CTE-W	0.140	0.18	+	–	–	+	–	ø	ø	ø	ø	–
2/9/00	UIS-CC	0.330	0.41	–	–	–	+	–	–	–	–	+	–
2/28/00	UIS-CC	0.200	0.25	–	–	–	+	–	–	–	–	+	–
2/29/00	UIS-CC	0.250	0.31	–	–	–	–	–	–	–	–	+	–
3/1/00	UIS-CC	0.250	0.31	–	–	–	–	–	–	–	–	+	–
3/30/00	UIS-CC	0.200	0.25	–	–	–	–	–	–	–	–	+	–
4/12/00	UIS-CC	0.250	0.31	–	–	–	–	–	–	–	–	+	–

Note: (+) Target organism DNA detected and confirmed; (–) Target organism DNA not detected; (ø) Sample not tested for this organism.

Table D-2. Phase 2 Membrane Dissolution PCR Presence/Absence Data Grouped by Sampling Sites, cont'd.

Sample Date	Sample Site	Sample Volume Filtered L	Effective Volume Tested mL	*Clostridium perfringens*	*C. rjejuni*	*H. pylori*	*Mycobacterium* *avium*	*Mycobacterium* *intracellulare*	*Salmonella*	STEC (Stx2)	*Yersinia enterocolitica*	*Aeromonas*	*Shigella*
4/24/00	UIS-CC	0.250	0.31	–	–	–	–	–	–	–	–	ø	–
5/25/00	UIS-CC	0.150	0.19	–	–	–	–	–	–	ø	ø	ø	–
6/2/00	UIS-CC	0.200	0.25	–	–	–	–	–	–	ø	ø	ø	–
6/12/00	UIS-CC	0.180	0.23	–	–	–	–	–	–	ø	ø	ø	–
7/12/00	UIS-CC	0.185	0.23	–	–	–	–	–	–	ø	ø	ø	–
8/28/00	UIS-CC	0.200	0.25	–	–	–	–	–	–	ø	ø	ø	–
10/3/00	UIS-CC	0.090	0.11	–	–	–	+	–	–	ø	ø	ø	–
10/16/00	UIS-CC	0.160	0.20	+	–	–	+	–	–	ø	ø	ø	–
11/1/00	UIS-CC	0.200	0.25	–	–	–	–	–	–	ø	ø	ø	–
12/5/00	UIS-CC	0.110	0.14	–	–	–	–	–	–	ø	ø	ø	–
1/2/01	UIS-CC	0.190	0.24	–	–	–	–	–	ø	ø	ø	ø	–
1/23/01	UIS-CC	0.150	0.19	–	–	–	–	–	ø	ø	ø	ø	–
2/7/01	UIS-CC	0.195	0.24	–	–	–	–	–	ø	ø	ø	ø	–
3/14/01	UIS-CC	0.115	0.14	–	–	–	–	–	ø	ø	ø	ø	–
3/19/01	UIS-CC	0.200	0.25	–	–	–	–	+	ø	ø	ø	ø	–
4/4/01	UIS-CC	0.200	0.25	–	–	–	+	–	ø	ø	ø	ø	–
5/8/01	UIS-CC	0.200	0.25	–	–	–	–	–	ø	ø	ø	ø	–
5/14/01	UIS-CC	0.200	0.25	–	–	–	–	–	ø	ø	ø	ø	–
2/15/00	UIS-SJC	0.170	0.21	–	–	–	+	–	–	–	–	+	–
2/22/00	UIS-SJC	0.200	0.25	–	ø	–	+	–	–	–	–	+	–
2/25/00	UIS-SJC	0.200	0.25	–	–	–	–	–	–	–	–	+	–
3/14/00	UIS-SJC	0.200	0.25	–	–	–	–	–	–	–	–	+	–
3/29/00	UIS-SJC	0.165	0.21	–	–	–	–	–	–	–	–	+	–
4/3/00	UIS-SJC	0.225	0.28	–	–	–	–	–	–	–	–	+	–
4/19/00	UIS-SJC	0.170	0.21	–	–	–	–	–	+	–	–	–	–
5/1/00	UIS-SJC	0.250	0.31	–	–	–	–	–	–	–	–	–	–

Note: (+) Target organism DNA detected and confirmed; (–) Target organism DNA not detected; (ø) Sample not tested for this organism.

Table D-2. Phase 2 Membrane Dissolution PCR Presence/Absence Data Grouped by Sampling Sites, cont'd.

Sample Date	Sample Site	Sample Volume Filtered L	Effective Volume Tested mL	*Clostridium perfringens*	*C. jejuni*	*H. pylori*	*Mycobacterium* *avium*	*Mycobacterium* *intracellulare*	*Salmonella*	STEC (Stx2)	*Yersinia enterocolitica*	*Aeromonas*	*Shigella*
5/24/00	UIS-SJC	0.350	0. 44	–	–	–	–	–	–	ø	ø	ø	–
6/5/00	UIS-SJC	0.300	0.38	–	–	–	–	–	–	ø	ø	ø	–
6/26/00	UIS-SJC	0.200	0.25	–	–	–	–	–		ø	ø	ø	–
7/27/00	UIS-SJC	0.190	0.24	–	–	–	–	–	–	ø	ø	ø	–
8/31/00	UIS-SJC	0.200	0.25	–	–	–	–	–	–	ø	ø	ø	–
9/5/00	UIS-SJC	0.200	0.25	–	–	–	–	+	–	ø	ø	ø	–
9/20/00	UIS-SJC	0.200	0.25	–	–	–	–	–	–	ø	ø	ø	–
10/5/00	UIS-SJC	0.150	0.19	–	–	–	–	–	–	ø	ø	ø	–
10/17/00	UIS-SJC	0.150	0.19	–	–	–	+	–	–	ø	ø	ø	–
10/26/00	UIS-SJC	0.200	0.25	–	–	–	–	–	–	ø	ø	ø	–
11/28/00	UIS-SJC	0.200	0.25	–	–	–	–	–	–	ø	ø	ø	–
12/6/00	UIS-SJC	0.155	0.19	–	–	–	–	–	–	ø	ø	ø	–
1/10/01	UIS-SJC	0.200	0.25	–	+	–	–	–	ø	ø	ø	ø	–
2/21/01	UIS-SJC	0.200	0.25	+	+	–	–	–	ø	ø	ø	ø	–
3/06/01	UIS-SJC	0.055	0.07	–	–	–	–	–	ø	ø	ø	ø	–
4/16/01	UIS-SJC	0.180	0.23	–	–	–	–	–	ø	ø	ø	ø	–
3/27/00	UIS-SRD	0.200	0.25	–	–	–	–	–	–	–	–	–	–
4/17/00	UIS-SRD	0.200	0.25	–	–	–	–	–	–	–	–	–	–
5/15/00	UIS-SRD	0.200	0.25	–	–	–	–	–	–	ø	ø	ø	ø
6/19/00	UIS-SRD	0.200	0.25	–	–	–	–	–	–	ø	ø	ø	ø
7/24/00	UIS-SRD	0.155	0.19	–	–	–	–	–	–	ø	ø	ø	ø
8/21/00	UIS-SRD	0.200	0.25	–	–	–	–	–	–	ø	ø	ø	ø
9/18/00	UIS-SRD	0.200	0.25	–	–	–	–	–	–	ø	ø	ø	ø
10/23/00	UIS-SRD	0.200	0.25	–	–	–	–	–	–	ø	ø	ø	ø
11/13/00	UIS-SRD	0.200	0.25	–	–	–	–	–	–	ø	ø	ø	ø

Note: (+) Target organism DNA detected and confirmed; (–) Target organism DNA not detected; (ø) Sample not tested for this organism.

Table D-2. Phase 2 Membrane Dissolution PCR Presence/Absence Data Grouped by Sampling Sites, cont'd.

Sample Date	Sample Site	Sample Volume Filtered L	Effective Volume Tested mL	*Clostridium perfringens*	*C. jejuni*	*H. pylori*	*Mycobacterium avium*	*Mycobacterium intracellulare*	*Salmonella*	STEC (Stx2)	*Yersinia enterocolitica*	*Aeromonas*	*Shigella*
12/18/00	UIS-SRD	0.150	0.19	–	–	–	–	–	–	ø	ø	ø	ø
1/22/01	UIS-SRD	0.200	0.25	–	–	–	–	–	ø	ø	ø	ø	ø
2/12/01	UIS-SRD	0.200	0.25	–	–	–	–	–	ø	ø	ø	ø	ø
3/23/00	UIS-SRU	0.250	0.31	–	–	–	–	–	–	–	–	–	–
4/13/00	UIS-SRU	0.200	0.25	–	+	–	–	–	–	–	–	–	–
5/11/00	UIS-SRU	0.200	0.25	–	–	–	–	–	–	–	–	–	–
6/15/00	UIS-SRU	0.250	0.31	–	–	–	–	–	–	ø	ø	ø	–
7/20/00	UIS-SRU	0.200	0.25	–	–	–	–	–	–	ø	ø	ø	–
8/17/00	UIS-SRU	0.200	0.25	–	–	–	–	–	–	ø	ø	ø	–
9/14/00	UIS-SRU	0.225	0.28	–	–	–	–	–	–	ø	ø	ø	–
10/19/00	UIS-SRU	0.200	0.25	–	–	–	–	–	–	ø	ø	ø	–
11/9/00	UIS-SRU	0.200	0.25	–	–	–	–	–	–	ø	ø	ø	–
12/14/00	UIS-SRU	0.200	0.25	–	–	–	–	–	–	ø	ø	ø	–
1/18/01	UIS-SRU	0.200	0.25	–	–	–	–	–	ø	ø	ø	ø	–
2/08/01	UIS-SRU	0.200	0.25	–	–	–	–	–	ø	ø	ø	ø	–
5/9/00	USW-B	0.200	0.25	–	–	–	–	–	–	–	–	–	–
8/7/00	USW-B	0.190	0.24	–	–	–	–	–	–	ø	ø	ø	–
12/19/00	USW-B	0.100	0.13	–	–	–	–	–	–	ø	ø	ø	–
7/12/01	USW-B	0.200	0.25	–	–	–	–	–	ø	ø	ø	ø	–
2/10/00	USW-C	0.500	0.63	–	–	–	–	–	–	–	–	–	–
2/14/00	USW-C	0.500	0.63	–	–	–	–	–	–	–	–	–	–
2/23/00	USW-C	0.500	0.63	–	–	–	–	–	–	–	–	–	–
2/24/00	USW-C	0.500	0.63	–	–	–	–	–	–	–	–	+	–
3/8/00	USW-C	0.600	0.75	–	–	–	–	–	–	–	–	–	–
3/9/00	USW-C	0.750	0.94	–	–	–	–	–	–	–	–	–	–

Note: (+) Target organism DNA detected and confirmed; (–) Target organism DNA not detected; (ø) Sample not tested for this organism.

Table D-2. Phase 2 Membrane Dissolution PCR Presence/Absence Data Grouped by Sampling Sites, cont'd.

Sample Date	Sample Site	Sample Volume Filtered L	Effective Volume Tested mL	*Clostridium perfringens*	*C. jejuni*	*H. pylori*	*Mycobacterium* *avium*	*Mycobacterium* *intracellulare*	*Salmonella*	STEC (Stx2)	*Yersinia enterocolitica*	*Aeromonas*	*Shigella*
3/20/00	USW-C	0.500	0.63	–	–	–	–	–	–	–	–	–	–
3/28/00	USW-C	0.500	0.63	–	–	–	–	–	–	–	–	–	–
4/11/00	USW-C	0.550	0.69	–	–	–	–	–	–	–	–	–	–
4/26/00	USW-C	0.500	0.63	–	–	–	–	–	–	–	–	ø	–
5/2/00	USW-C	0.550	0.69	–	–	–	–	–	–	–	–	ø	–
5/3/00	USW-C	0.500	0.63	–	–	–	–	–	–	–	–	ø	–
5/18/00	USW-C	0.700	0.88	–	–	–	–	–	–	ø	ø	ø	–
5/30/00	USW-C	0.650	0.81	–	–	–	–	–	–	ø	ø	ø	–
6/14/00	USW-C	0.600	0.75	–	–	–	–	–	–	ø	ø	ø	–
6/21/00	USW-C	0.500	0.63	–	–	–	–	–	–	ø	ø	ø	–
7/17/00	USW-C	0.300	0.38	–	–	–	–	–	–	ø	ø	ø	–
7/26/00	USW-C	0.300	0.38	–	–	–	–	–	–	ø	ø	ø	–
9/28/00	USW-C	0.320	0.40	–	–	–	–	–	–	ø	ø	ø	–
10/2/00	USW-C	0.325	0.41	–	–	–	–	–	–	ø	ø	ø	–
10/4/00	USW-C	0.325	0.41	–	–	–	–	–	–	ø	ø	ø	–
10/25/00	USW-C	0.250	0.31	–	–	–	–	–	–	ø	ø	ø	–
11/30/00	USW-C	0.150	0.19	–	–	–	–	–	–	ø	ø	ø	–
12/4/00	USW-C	0.300	0.38	–	–	–	–	–	–	ø	ø	ø	–
1/4/01	USW-C	0.265	0.33	–	–	–	–	–	ø	ø	ø	ø	–
1/29/01	USW-C	0.325	0.41	–	–	–	–	–	ø	ø	ø	ø	–
1/31/01	USW-C	0.300	0.38	–	–	–	–	–	ø	ø	ø	ø	–
2/5/01	USW-C	0.275	0.34	–	–	–	–	–	ø	ø	ø	ø	–
3/20/01	USW-C	0.250	0.31	–	–	–	–	–	ø	ø	ø	ø	–
4/19/01	USW-C	0.225	0.28	–	–	–	–	–	ø	ø	ø	ø	–
5/9/01	USW-C	0.300	0.38	–	–	–	–	–	ø	ø	ø	ø	–
6/5/01	USW-C	0.300	0.38	–	–	–	–	–	ø	ø	ø	ø	
6/13/01	USW-C	0.275	0.34	–	–	–	–	–	ø	ø	ø	ø	–

Note: (+) Target organism DNA detected and confirmed; (–) Target organism DNA not detected; (ø) Sample not tested for this organism

Table D-2. Phase 2 Membrane Dissolution PCR Presence/Absence Data Grouped by Sampling Sites, cont'd.

Sample Date	Sample Site	Sample Volume Filtered L	Effective Volume Tested mL	*Clostridium perfringens*	*C. jejuni*	*H. pylori*	*Mycobacterium* *avium*	*Mycobacterium* *intracellulare*	*Salmonella*	STEC (Stx2)	*Yersinia enterocolitica*	*Aeromonas*	*Shigella*
6/19/01	USW-C	0.275	0.34	–	–	–	–	–	ø	ø	ø	ø	–
6/26/01	USW-C	0.300	0.38	–	–	–	–	–	ø	ø	ø	ø	–
7/10/01	USW-C	0.300	0.38	–	–	–	–	–	ø	ø	ø	ø	–
4/27/00	USW-H	0.500	0.63	–	–	–	–	–	–	–	–	–	–
6/28/00	USW-H	0.400	0.50	–	–	–	–	–	–	ø	ø	ø	–
8/29/00	USW-H	0.500	0.63	–	–	–	–	–	–	ø	ø	ø	–
5/30/01	USW-H	0.300	0.38	–	–	–	–	–	ø	ø	ø	ø	–
11/6/00	USW-M	0.200	0.25	–	–	–	–	–	–	ø	ø	ø	–
1/8/01	USW-M	0.300	0.38	–	–	–	–	–	ø	ø	ø	ø	–
3/27/01	USW-M	0.275	0.34	–	–	–	–	–	ø	ø	ø	ø	–
5/14/01	USW-M	0.300	0.38	–	–	–	–	–	ø	ø	ø	ø	–
4/18/00	USW-O	0.200	0.25	–	–	–	–	–	–	–	–	–	–
6/13/00	USW-O	0.250	0.31	–	–	–	–	–	–	ø	ø	ø	–
9/12/00	USW-O	0.300	0.38	–	–	–	–	–	–	ø	ø	ø	–
2/27/01	USW-O	0.100	0.13	–	–	–	–	–	ø	ø	ø	ø	–
2/7/00	USW-P	0.400	0.50	–	–	–	–	–	–	–	–	+	–
2/8/00	USW-P	0.350	0.44	–	–	–	+	–	–	–	–	+	–
2/16/00	USW-P	0.300	0.38	–	–	–	+	–	–	–	–	+	–
2/17/00	USW-P	0.300	0.38	–	–	–	–	–	–	–	–	+	–
3/2/00	USW-P	0.300	0.38	–	–	–	–	–	–	–	–	+	–
3/7/00	USW-P	0.250	0.31	–	–	–	–	–	–	–	–	+	–
3/13/00	USW-P	0.200	0.25	–	–	–	–	–	–	–	–	+	–
4/4/00	USW-P	0.215	0.27	–	–	–	–	–	–	–	–	+	–
4/5/00	USW-P	0.200	0.25	–	–	–	–	–	–	–	–	+	–
4/6/00	USW-P	0.200	0.25	–	–	–	–	–	–	–	–	+	–

Note: (+) Target organism DNA detected and confirmed; (–) Target organism DNA not detected; (ø) Sample not tested for this organism

Table D-2. Phase 2 Membrane Dissolution PCR Presence/Absence Data Grouped by Sampling Sites, cont'd.

Sample Date	Sample Site	Sample Volume Filtered L	Effective Volume Tested mL	*Clostridium perfringens*	*C. jejuni*	*H. pylori*	*Mycobacterium* *avium*	*Mycobacterium* *intracellulare*	*Salmonella*	STEC (Stx2)	*Yersinia enterocolitica*	*Aeromonas*	*Shigella*
5/22/00	USW-P	0.300	0.38	–	–	–	–	–	–	ø	ø	ø	–
5/31/00	USW-P	0.300	0.38	–	–	–	–	–	–	ø	ø	ø	–
6/1/00	USW-P	0.300	0.38	–	–	–	–	–	–	ø	ø	ø	–
6/7/00	USW-P	0.325	0.21	–	–	–	–	–	–	ø	ø	ø	–
6/8/00	USWP	0.350	0.44	–	–	–	–	–	–	ø	ø	ø	–
6/20/00	USW-P	0.200	0.25	–	–	–	–	–	–	ø	ø	ø	–
6/27/00	USW-P	0.150	0.19	–	–	–	–	–	–	ø	ø	ø	–
7/10/00	USW-P	0.200	0.25	–	–	–	–	–	–	ø	ø	ø	–
7/19/00	USW-P			Sample lost – Lab Accident									
8/24/00	USW-P	0.200	0.25	–	–	–	–	–	–	ø	ø	ø	–
9/21/00	USW-P	0.275	0.34	–	–	–	–	–	–	ø	ø	ø	–
10/10/00	USW-P	0.205	0.26	–	–	–	–	–	–	ø	ø	ø	–
10/18/00	USW-P	0.225	0.28	–	–	–	–	–	–	ø	ø	ø	–
10/30/00	USW-P	0.200	0.25	–	–	–	–	–	–	ø	ø	ø	–
11/15/00	USW-P	0.200	0.25	–	–	–	–	–	–	ø	ø	ø	–
12/11/00	USW-P	0.200	0.25	–	–	–	–	–	–	ø	ø	ø	–
1/16/01	USW-P	0.225	0.28	–	–	–	–	–	ø	ø	ø	ø	–
1/24/01	USW-P	0.200	0.25	–	–	–	–	–	ø	ø	ø	ø	–
2/13/01	USW-P	0.200	0.25	–	–	–	–	–	ø	ø	ø	ø	–
2/26/01	USW-P	0.200	0.28	–	–	–	–	–	ø	ø	ø	ø	–
3/7/01	USW-P	0.200	0.25	–	–	–	–	–	ø	ø	ø	ø	–
3/13/01	USW-P	0.225	0.28	–	–	–	–	–	ø	ø	ø	ø	–
4/5/01	USW-P	0.170	0.21	–	–	–	–	–	ø	ø	ø	ø	–
4/24/01	USW-P	0.200	0.25	–	–	–	–	–	ø	ø	ø	ø	–
4/30/01	USW-P	0.200	0.25	–	–	–	–	–	ø	ø	ø	ø	–
5/7/01	USW-P	0.200	0.25	–	–	–	–	–	ø	ø	ø	ø	–

Note: (+) Target organism DNA detected and confirmed; (–) Target organism DNA not detected; (ø) Sample not tested for this organism.

Table D-2. Phase 2 Membrane Dissolution PCR Presence/Absence Data Grouped by Sampling Sites, cont'd.

Sample Date	Sample Site	Sample Volume Filtered L	Effective Volume Tested mL	*Clostridium perfringens*	*C. jejuni*	*H. pylori*	*Mycobacterium avium*	*Mycobacterium intracellulare*	*Salmonella*	STEC (Stx2)	*Yersinia enterocolitica*	*Aeromonas*	*Shigella*
3/16/00	USW-R	0.450	0.56	–	–	–	–	–	–	–	–	+	–
4/20/00	USW-R	0.400	0.50	–	–	–	–	–	–	–	–	–	–
7/20/00	USW-R	0.400	0.50	–	–	–	–	–	–	ø	ø	ø	–
11/16/00	USW-R	0.300	0.38	–	–	–	–	–	–	ø	ø	ø	–
3/15/00	USW-S	0.165	0.21	–	–	–	–	–	–	–	–	+	–
5/11/00	USW-S	0.200	0.25	–	–	–	–	–	–	–	–	–	–
10/11/00	USW-S	0.200	0.25	–	–	–	–	–	–	ø	ø	ø	
6/20/01	USW-S	0.190	0.24	–	–	–	–	–	ø	ø	ø	ø	–
5/9/00	USW-T	0.175	0.22	–	–	–	–	–	–	–	–	–	–
9/26/00	USW-T	0.105	0.13	–	–	–	–	–	–	ø	ø	ø	–
1/30/01	USW-T	0.040	0.05	–	–	–	–	–	ø	ø	ø	ø	–
5/29/01	USW-T	0.110	0.14	–	–	–	–	–	ø	ø	ø	ø	–
5/9/00	USW-V	0.200	0.25	–	–	–	–	–	–	–	–	–	–
8/7/00	USW-V	0.125	0.16	–	–	–	–	–	–	ø	ø	ø	–
12/19/00	USW-V	0.200	0.25	–	–	–	–	–	–	ø	ø	ø	–
7/12/01	USW-V	0.200	0.25	–	–	–	–	–	ø	ø	ø	ø	–
7/24/00	USW-W	0.200	0.25	–	–	–	–	–	–	ø	ø	ø	–
9/19/00	USW-W	0.200	0.25	–	–	–	–	–	–	ø	ø	ø	–
12/18/00	USW-W	0.225	0.25	–	–	–	–	–	–	ø	ø	ø	–
5/21/01	USW-W	0.200	0.25	–	–	–	+	–	ø	ø	ø	ø	–
5/23/00	USW-Y	0.300	0.38	–	–	–	–	–	–	–	–	–	–
7/17/00	USW-Y	0.200	0.25	–	–	–	–	–	ø	ø	ø	ø	–
10/30/00	USW-Y	0.200	0.25	–	–	–	–	–	ø	ø	ø	ø	–
4/02/01	USW-Y	0.180	0.23	–	–	–	+	–	ø	ø	ø	ø	–

Note: (+) Target organism DNA detected and confirmed; (–) Target organism DNA not detected; (ø) Sample not tested for this organism.

Table D-3. 10-L Nonselective Enrichment Chlorinated Effluent Samples-PCR Presence/Absence Results

Sample Date	Sample Site	*Salmonella*	STEC (Stx2)	*Yersinia enterocolitica*	*Aeromonas*	*Shigella*
6/27/00	CSE-J	–	–	–	+	–
9/7/00	CSE-J	+	–	–	+	–
1/8/01	CSE-J	–	–	–	+	–
7/18/01	CSE-J	+	–	–	+	–
4/25/00	CSE-L	+	–	–	+	–
6/6/00	CSE-L	+	–	–	+	–
9/26/00	CSE-L	–	–	–	+	–
2/6/01	CSE-L	–	–	–	+	–
3/6/00	CTE-S	+	–	–	+	–
3/22/00	CTE-S	–	–	–	+	–
7/11/00	CTE-S	–	–	–	+	–
11/2/00	CTE-S	–	–	–	–	–
3/21/00	CTE-W	+	–	–	+	–
4/10/00	CTE-W	+	–	–	+	–
9/6/00	CTE-W	–	–	–	+	–
2/15/01	CTE-W	–	–	–	+	–

Table D-3. 10-L Nonselective Enrichment Urban Runoff Affected Stream Samples-PCR Presence/Absence Results

Sample Date	Sample Site	*Salmonella*	STEC (Stx2)	*Yersinia enterocolitica*	*Aeromonas*	*Shigella*
2/9/00	UIS-CC	+	+	–	–	–
2/28/00	UIS-CC	+	+	–	+	–
2/29/00	UIS-CC	+	+	–	+	–
3/1/00	UIS-CC	+	–	–	+	–
3/30/00	UIS-CC	+	–	–	+	–
4/12/00	UIS-CC	+	–	–	+	–
4/24/00	UIS-CC	+	+	–	+	–
5/25/00	UIS-CC	+	–	–	+	–
6/2/00	UIS-CC	+	–	–	+	–
6/12/00	UIS-CC	+	–	–	+	–
7/12/00	UIS-CC	+	–	–	+	–
8/28/00	UIS-CC	+	+	–	+	–
10/3/00	UIS-CC	+	–	–	+	–
10/16/00	UIS-CC	+	+	–	+	–
11/1/00	UIS-CC	+	+	–	+	–
12/5/00	UIS-CC	+	+	–	+	–
1/2/01	UIS-CC	+	+	–	+	–
1/23/01	UIS-CC	+	–	–	+	–
2/7/01	UIS-CC	–	–	–	+	–
3/14/01	UIS-CC	+	–	–	+	–
3/19/01	UIS-CC	+	–	–	+	–
4/4/01	UIS-CC	+	+	–	+	–
5/8/01	UIS-CC	+	–	–	+	–
5/14/01	UIS-CC	+	–	–	+	–
2/15/00	UIS-SJC	+	+	–	+	–
2/22/00	UIS-SJC	+	+	–	+	–
2/25/00	UIS-SJC	+	+	–	+	–
3/14/00	UIS-SJC	+	+	–	+	–
3/29/00	UIS-SJC	+	+	–	+	–
4/3/00	UIS-SJC	–	–	–	–	–
4/19/00	UIS-SJC	+	+	–	+	–
5/1/00	UIS-SJC	+	–	–	+	–
5/24/00	UIS-SJC	+	–	–	+	–
6/5/00	UIS-SJC	+	–	–	+	–
6/26/00	UIS-SJC	+	+	–	+	–
7/27/00	UIS-SJC	+	+	–	+	–
8/31/00	UIS-SJC	+	+	–	+	–
9/5/00	UIS-SJC	+	–	–	+	–

Table D-3. 10-L Nonselective Enrichment Urban Runoff Impacted Stream Samples-PCR Presence/Absence Results, cont'd.

Sample Date	Sample Site	*Salmonella*	STEC (Stx2)	*Yersinia enterocolitica*	*Aeromonas*	*Shigella*
9/20/00	UIS-SJC	+	+	–	+	–
10/5/00	UIS-SJC	+	–	–	+	–
10/17/00	UIS-SJC	+	+	+	–	–
10/26/00	UIS-SJC	+	+	–	+	–
11/28/00	UIS-SJC	+	+	–	+	–
12/6/00	UIS-SJC	+	+	–	+	–
1/10/01	UIS-SJC	–	+	–	+	–
2/21/01	UIS-SJC	+	–	–	+	–
3/06/01	UIS-SJC	+	–	–	+	–
4/16/01	UIS-SJC	+	+	–	+	–
3/27/00	UIS-SRD	+	+	–	+	–
4/17/00	UIS-SRD	+	+	–	+	–
5/15/00	UIS-SRD	+	–	–	–	–
6/19/00	UIS-SRD	+	+	–	+	–
7/24/00	UIS-SRD	+	+	–	+	–
8/21/00	UIS-SRD	+	+	–	+	–
9/18/00	UIS-SRD	+	+	–	+	–
10/23/00	UIS-SRD	+	+	–	–	–
11/13/00	UIS-SRD	+	+	–	+	–
12/18/00	UIS-SRD	+	+	–	+	–
1/22/01	UIS-SRD	+	+	–	+	–
2/12/01	UIS-SRD	–	+	–	+	–
3/23/00	UIS-SRU	+	+	–	+	–
4/13/00	UIS-SRU	+	–	–	+	–
5/11/00	UIS-SRU	+	–	–	+	–
6/15/00	UIS-SRU	+	+	–	+	–
7/20/00	UIS-SRU	+	+	–	+	–
8/17/00	UIS-SRU	+	–	–	+	–
9/14/00	UIS-SRU	+	+	–	+	–
10/19/00	UIS-SRU	+	+	–	+	–
11/9/00	UIS-SRU	+	+	–	+	–
12/14/00	UIS-SRU	+	+	–	+	–
1/18/01	UIS-SRU	+	+	–	+	–
2/08/01	UIS-SRU	–	+	–	+	–

Table D-3. 10-L Nonselective Enrichment Untreated Surface Water Samples-PCR Presence/Absence Results

Sample Date	Sample Site	*Salmonella*	STEC (Stx2)	*Yersinia enterocolitica*	*Aeromonas*	*Shigella*
5/9/00	USW-B	+	–	–	+	–
8/7/00	USW-B	+	–	–	+	–
12/19/00	USW-B	+	+	–	+	–
7/12/01	USW-B	–	–	–	+	–
2/10/00	USW-C	+	–	–	+	–
2/14/00	USW-C	+	+	–	+	–
2/23/00	USW-C	+	+	–	+	–
2/24/00	USW-C	+	+	–	+	–
3/8/00	USW-C	+	–	–	+	–
3/9/00	USW-C	+	–	–	+	–
3/20/00	USW-C	+	+	–	+	–
3/28/00	USW-C	+	–	–	–	–
4/11/00	USW-C	+	–	–	+	–
4/26/00	USW-C	+	–	–	+	–
5/2/00	USW-C	+	+	–	+	–
5/3/00	USW-C	+	+	–	+	–
5/18/00	USW-C	+	–	–	+	–
5/30/00	USW-C	+	–	–	+	–
6/14/00	USW-C	+	–	–	+	–
6/21/00	USW-C	+	–	–	+	–
7/17/00	USW-C	+	–	–	+	–
7/26/00	USW-C	–	–	–	+	–
9/28/00	USW-C	–	–	–	+	–
10/2/00	USW-C	–	–	–	+	–
10/4/00	USW-C	–	–	–	–	–
10/25/00	USW-C	–	–	–	–	–
11/30/00	USW-C	+	–	–	–	–
12/4/00	USW-C	–	–	–	–	–
1/4/01	USW-C	–	+	–	+	–
1/29/01	USW-C	–	–	–	–	–
1/31/01	USW-C	–	–	–	+	–
2/5/01	USW-C	–	–	–	–	–
3/20/01	USW-C	+	–	–	+	–
4/19/01	USW-C	+	–	–	+	–
5/9/01	USW-C	–	–	–	+	–
6/5/01	USW-C	–	–	–	+	–
6/13/01	USW-C	–	–	–	+	–
6/19/01	USW-C	+	–	–	+	–
6/26/01	USW-C	–	–	–	+	–
7/10/01	USW-C	–	–	–	+	–

Table D-3. 10-L Nonselective Enrichment Untreated Surface Water Samples-PCR Presence/Absence Results, cont'd

Sample Date	Sample Site	*Salmonella*	STEC (Stx2)	*Yersinia enterocolitica*	*Aeromonas*	*Shigella*
4/27/00	USW-H	+	–	–	+	–
6/28/00	USW-H	+	–	–	+	–
8/29/00	USW-H	+	–	–	+	–
5/30/01	USW-H	–	–	–	–	–
11/6/00	USW-M	–	–	–	–	–
1/8/01	USW-M	–	+	–	+	–
3/27/01	USW-M	+	–	–	+	–
5/14/01	USW-M	+	–	–	+	–
4/18/00	USW-O	+	+	–	+	–
6/13/00	USW-O	+	–	–	+	–
9/12/00	USW-O	+	+	–	+	–
2/27/01	USW-O	+	–	–	+	–
2/7/00	USW-P	+	–	–	+	–
2/8/00	USW-P	+	–	–	+	–
2/16/00	USW-P	+	–	–	+	–
2/17/00	USW-P	+	–	–	+	–
3/2/00	USW-P	+	–	–	+	–
3/7/00	USW-P	+	–	–	+	–
3/13/00	USW-P	+	–	–	+	–
4/4/00	USW-P	+	+	–	+	–
4/5/00	USW-P	+	–	–	+	–
4/6/00	USW-P	+	–	–	+	–
5/22/00	USW-P	–	–	–	+	–
5/31/00	USW-P	–	–	–	+	–
6/1/00	USW-P	+	–	–	+	–
6/7/00	USW-P	+	–	–	+	–
6/8/00	USW-P	+	–	–	+	–
6/20/00	USW-P	+	–	–	+	–
6/27/00	USW-P	+	–	–	+	–
7/10/00	USW-P	–	–	–	+	–
7/19/00	USW-P	–	–	–	+	–
8/24/00	USW-P	–	–	–	+	–
9/21/00	USW-P	–	–	–	+	–
10/10/00	USW-P	–	–	–	–	–
10/18/00	USW-P	+	–	–	+	–
10/30/00	USW-P	–	–	–	+	–
11/15/00	USW-P	–	–	–	+	–

Table D-3. 10-L Nonselective Enrichment Untreated Surface Water Samples-PCR Presence/Absence Results, cont'd.

Sample Date	Sample Site	*Salmonella*	STEC (Stx2)	*Yersinia enterocolitica*	*Aeromonas*	*Shigella*
12/11/00	USW-P	–	–	–	+	–
1/16/01	USW-P	–	–	–	+	–
1/24/01	USW-P	–	–	–	+	–
2/13/01	USW-P	+	–	–	+	–
2/26/01	USW-P	+	–	–	+	–
3/7/01	USW-P	+	–	–	+	–
3/13/01	USW-P	+	–	–	+	–
4/5/01	USW-P	+	–	–	+	–
4/24/01	USW-P	–	–	–	+	–
4/30/01	USW-P	–	–	–	+	–
5/7/01	USW-P	–	–	–	+	–
3/16/00	USW-R	+	+	–	+	–
4/20/00	USW-R	+	–	–	+	–
7/20/00	USW-R	+	–	–	+	–
11/16/00	USW-R	+	+	–	+	–
5/11/00	USW-S	+	+	–	+	–
7/12/00	USW-S	–	–	–	+	–
10/11/00	USW-S	–	–	–	–	–
6/20/01	USW-S	+	–	–	–	–
5/9/00	USW-T	+	–	–	+	–
9/26/00	USW-T	+	–	–	+	–
1/30/01	USW-T	–	+	–	+	–
5/29/01	USW-T	+	–	–	+	–
5/9/00	USW-V	+	–	–	+	–
8/7/00	USW-V	–	+	–	+	–
12/19/00	USW-V	+	+	–	+	–
7/12/01	USW-V	–	–	–	+	–
7/24/00	USW-W	+	+	–	+	–
9/19/00	USW-W	–	+	+	+	–
12/18/00	USW-W	+	+	–	+	–
5/21/01	USW-W	–	+	–	–	–
5/23/00	USW-Y	+	–	–	+	–
7/17/00	USW-Y	+	–	–	+	–
10/30/00	USW-Y	–	–	–	+	–
4/02/01	USW-Y	–	–	+	+	–

APPENDIX E

Quality Assurance – Indicator Organism Duplicate Analyses

Table E-1. Quality Assurance – Indicator Organism Duplicate Analyses.

Sample Date	Sample Site	Replicate	Total coliform MPN/100mL	Fecal coliform MPN/100mL	*E.coli* MPN/100mL	Fecal Strep MPN/100mL	Enterococcus MPN/100mL	*Clostridium perfringens* CFU/100ml	CN13 Somatic MPN/100mL	F-amp male specific MPN/100mL
1/24/01	USW-P	1	17	4	2	2	2	2.7	30	13
		2	30	11	11	9	9	1	27	4
		Mean	23	7	7	6	6	2	30	9
1/29/01	USW-C	1	2	2	2	8	4	0.6	7	<2
		2	11	4	4	2	2	1	9	2
		Mean	7	3	3	5	3	0.8	8	2
1/30/01	USW-T	1	70,000	11,000	11,000	23,000	390	410	2,200	170
		2	90,000	22,000	22,000	13,000	560	460	1,600	220
		Mean	80,000	17,000	17,000	18,000	480	440	1,800	200
2/5/01	USW-C	1	13	13	13	11	2	1	8	4
		2	8	4	4	13	4	0.6	13	<2
		Mean	10	8	8	12	3	0.8	11	3
2/6/01	CSE-L	1	4	<2	<2	<2	<2	6	<2	<2
		2	7	<2	<2	<2	<2	12	<2	<2
		Mean	6	<2	<2	<2	<2	9	<2	<2
2/7/01	UIS-CC	1	30,000	13,000	13,000	17,000	17,000	76	300	30
		2	30,000	17,000	17,000	30,000	30,000	88	130	170
		Mean	30,000	15,000	15,000	24,000	24,000	77	220	100
2/8/01	UIS-SRU	1	8,000	300	170	2,300	2,300	14	30	14
		2	5,000	280	280	500	500	15	NO DATA	NO DATA
		Mean	6,500	290	220	1,400	1,400	14		
2/12/01	UIS-SRD	1	30,000	5,000	5,000	2,300	2,300	49	9,000	500
		2	4,300	1,300	1,300	2,200	1,700	30	16,000	80
		Mean	17,000	2,200	2,200	2,250	2,000	40	13,000	290

Table E-1. Quality Assurance – Indicator Organism Duplicate Analyses, cont'd.

Sample Date	Sample Site	Replicate	Total coliform MPN/100mL	Fecal coliform MPN/100mL	*E.coli* MPN/100mL	Fecal Strep MPN/100mL	Enterococcus MPN/100mL	*Clostridium perfringens* CFU/100ml	CN13 Somatic MPN/100mL	F-amp male specific MPN/100mL
2/13/01	USW-P	1	22	14	11	13	13	15	4	<2
		2	17	8	4	8	4	15	<2	<2
		Mean	20	11	8	10	8	15	3	<2
2/15/01	CTE-W	1	<2	<2	<2	<2	<2	3	14	170
		2	2	<2	<2	<2	<2	6.3	8	170
		Mean	2	<2	<2	<2	<2	5	11	170
2/27/01	USW-O	1	220	50	30	14	7	69	500	13
		2	350	30	30	50	8	77	170	13
		Mean	290	40	30	32	8	73	340	13
3/6/01	UIS-SJC	1	170,000	30,000	30,000	300,000	130,000	920	1,700	240
		2	240,000	24,000	13,000	50,000	17,000	980	3,000	500
		Mean	200,000	27,000	22,000	180,000	74,000	950	2,400	370
3/20/01	USW-C	1	130	22	22	17	11	2	170	23
		2	330	80	80	30	11	2	130	32
		Mean	230	51	51	24	11	2	150	28
3/28/01	USW-M	1	90	30	17	8	8	0.2	80	17
		2	30	17	11	23	23	0.5	30	23
		Mean	60	24	14	16	16	0.4	60	20
4/2/01	USW-Y	1	900	80	80	220	9	12	4	2
		2	3,000	70	70	230	11	9	<2	<2
		Mean	2,000	75	75	230	10	10	3	2
4/4/01	UIS-CC	1	13,000	1,400	1,400	300	300	62	300	27
		2	17,000	1,100	1,100	130	80	90	1,300	140
		Mean	15,000	1,300	1,300	220	190	76	800	84
4/16/01	UIS-SJC	1	13,000	700	300	700	120	19	30	130
		2	1,300	230	50	800	230	18	30	130
		Mean	7,200	470	180	750	180	19	30	130

Table E-1. Quality Assurance – Indicator Organism Duplicate Analyses, cont'd.

Sample Date	Sample Site	Replicate	Total coliform MPN/100mL	Fecal coliform MPN/100mL	*E.coli* MPN/100mL	Fecal Strep MPN/100mL	Enterococcus MPN/100mL	*Clostridium perfringens* CFU/100ml	CN13 Somatic MPN/100mL	F-amp male specific MPN/100mL
4/19/01	USW-C	1	280	7	7	<2	<2	0.6	23	8
		2	230	23	23	<2	<2	0.6	13	2
		Mean	260	15	15	<2	<2	0.6	18	5
4/24/01	USW-P	1	8	4	4	<2	<2	9	<2	<2
		2	14	2	2	<2	<2	10	<2	<2
		Mean	11	3	3	<2	<2	10	<2	<2
4/30/01	USW-P	1	700	2	2	<2	<2	1	<2	<2
		2	210	<2	<2	<2	<2	4	2	<2
		Mean	460	2	2	<2	<2	3	2	<2
5/7/01	USW-P	1	<2	<2	<2	<2	<2	1	<2	<2
		2	<2	<2	<2	<2	<2	2	<2	<2
		Mean	<2	<2	<2	<2	<2	2	<2	<2
5/8/01	UIS-CC	1	11,000	700	700	300	130	30	500	220
		2	8,000	300	300	900	300	29	130	30
		Mean	9,500	500	500	600	220	30	320	130
5/15/01	USW-M	1	11	8	8	<2	<2	<0.1	23	2
		2	13	4	<2	2	<2	<0.1	14	4
		Mean	12	6	5	2	<2	<0.1	19	3
5/21/01	USW-W	1	22	4	4	7	2	1	<2	<2
		2	70	7	7	4	2	0.6	4	<2
		Mean	46	6	6	6	2	0.8	3	<2
5/29/01	USW-T	1	35,000	5,000	800	900	70	38	80	170
		2	35,000	3,000	800	2,300	40	42	170	220
		Mean	35,000	4,000	800	1,600	60	40	120	200
5/30/01	USW-H	1	70	<2	<2	2	2	0.2	<2	<2
		2	80	2	2	2	2	0.2	<2	<2
		Mean	75	2	2	2	2	0.2	<2	<2

APPENDIX F

TECHNOLOGY TRANSFER – TWO LAB SPLIT SAMPLES

Table F-1. Samples Concentrated, Enriched and Extracted at MWD Lab, and Split with LACSD Lab for Analysis.

		Sample				
		A	B	C	D	E
Organism	Test	CSD/MWD	CSD/MWD	CSD/MWD	CSD/MWD	CSD/MWD
Salmonella	PCR	+/+	–/+	–/+	+/+	+/+
	Probe	+/+	–/+	–/+	–/+	+/+
STEC	PCR	–/–	–/–	–/+	–/–	+/+
	Probe	–/–	–/–	–/+	–/–	+/+
Yersinia enterocolitica	PCR	–/–	–/–	–/–	–/–	–/–
	Probe	–/–	–/–	–/–	–/–	–/–
Aeromonas hydrophila	PCR	+/+	+/+	+/+	+/+	+/+
	Probe	–/+	–/+	–/+	–/+	–/+
Shigella	PCR	–/–	–/–	–/–	–/–	–/–
	Probe	–/–	–/–	–/–	–/–	–/–

Note: In addition to the five samples listed in Table F-1, one ultapure laboratory water sample was also concentrated and tested as a negative control. All tests were negative on the blank sample at both laboratories.

Table F-2. Samples Concentrated at LACSD Marine Lab, Enriched and Extracted at LACSD Project Lab, and Split with MWD Lab for Analysis.

		Sample				
		A	B	C	D	E
Organism	Test	CSD/MWD	CSD/MWD	CSD/MWD	CSD/MWD	CSD/MWD
Salmonella	PCR	+/+	–/–	+/+	+/+	+/+
	Probe	–/+	–/–	+/+	+/+	+/+
STEC	PCR	+/–	–/–	+/–	–/–	+/+
	Probe	–/–	–/–	+/–	–/–	+/+
Yersinia	PCR	–/–	–/–	–/–	–/–	–/–
enterocolitica	Probe	–/–	–/–	–/–	–/–	–/–
Aeromonas	PCR	–/+	–/+	–/+	+/+	+/+
hydrophila	Probe	–/+	–/+	–/+	–/+	–/+
Shigella	PCR	–/–	–/–	–/–	–/–	–/–
	Probe	–/–	–/–	–/–	–/–	–/–

Note: In addition to the five samples listed in Table F-2, one ultapure laboratory water sample was also concentrated and tested as a negative control. All tests were negative on the blank sample at both laboratories.

GLOSSARY

Amplicon - A term for any small, replicating DNA fragment.

Complementary DNA (cDNA) - single-stranded DNA that is complementary to messenger RNA or DNA that has been synthesized from messenger RNA by reverse transcriptase; the single- stranded form is often used as a probe in physical mapping.

DNA - 1) An organic acid and polymer composed of four nitrogenous bases – adenine, thymine, cytosine, and guanine linked via intervening units of phosphate and the pentose sugar deoxyribose. DNA is the genetic material of most organisms and usually exists as a double-stranded molecule in which two antiparallel strands are held together by hydrogen bonds between adeninethymine and cytosine-guanine. 2) Deoxyribonucleic acid, the material which carries the hereditary instructions (the "blueprint") that determine the formation of all living organisms. Genes, the organizers of inheritance, are composed of DNA.

Electrophoresis - A method of separating large molecules (such as DNA fragments or proteins) from a mixture of similar molecules. An electric current is passed through a medium containing the mixture, and each kind of molecule travels through the medium at a different rate, depending on its electrical charge and size. Separation is based on these differences. Agarose and acrylamide gels are the media commonly used for electrophoresis of proteins and nucleic acids. Specific electrophoretic methods include Northern blot, Southern blot , and Western blot.

Florophore - A protein which fluoresces or glows green visible light when excited by UV light with a wavelength of 395 nanometers. It can function as a biological marker when attached to other proteins.

Homology - Similarity in DNA or protein sequences between individuals of the same species or among different species. A similarity due to inheritance from a common ancestor.

Hybridization - The process of joining two complementary strands of DNA or one each of DNA and RNA to form a double-stranded molecule.

Messenger RNA (mRNA) - The class of RNA molecules that copies the genetic information from DNA, in the nucleus, and carries it to ribosomes, in the cytoplasm, where it is translated into protein.

Microarray - Sets of miniaturized chemical reaction areas that may also be used to test DNA fragments, antibodies, or proteins.

Oligonucleotide - A DNA polymer composed of only a few nucleotides. (See Nucleotide.) A molecule usually composed of 25 or fewer nucleotides, used as a DNA synthesis primer.

Phylogenetic - of or relating to the evolutionary development of organisms, "phylogenetic development."

PCR - A method for amplifying a DNA base sequence using a heat-stable polymerase and two 20-base primers , one complementary to the (+) -strand at one end of the sequence to be amplified and the other complementary to the (-) -strand at the other end. Because the newly synthesized DNA strands can subsequently serve as additional templates for the same primer sequences, successive rounds of primer annealing, strand elongation, and dissociation produce rapid and highly specific amplification of the desired sequence. PCR also can be used to detect the existence of the defined sequence in a DNA sample.

PCR Primer - A short DNA sequence homologous to a region on the DNA to be amplified. The primer provides the borders for the region of DNA to be amplified.

Polymerase - an enzyme that catalyzes the formation of new DNA and RNA from an existing strand of DNA or RNA

Probe - A single-stranded nucleic acid molecule with a known nucleotide sequence, which is labeled in some way (for example, radioactively, fluorescently, or immunologically) and used to find and mark certain DNA or RNA sequences of interest to a researcher by hybridizing to it.

Reverse transcriptase - An enzyme isolated from retrovirus-infected cells that synthesizes a complementary (c)DNA strand from an RNA template.

Ribosomes - Small cellular components composed of specialized ribosomal RNA and protein; site of protein synthesis.

Ribosomal RNA (rRNA) - A class of RNA found in the ribosomes of cells.

RNA - A chemical found in the nucleus and cytoplasm of cells; it plays an important role in protein synthesis and other chemical activities of the cell. The structure of RNA is similar to that of DNA. There are several classes of RNA molecules, including messenger RNA, transfer RNA, ribosomal RNA, and other small RNAs, each serving a different purpose. RNA is the genetic material of some viruses.

Shiga toxin - Group of structurally related toxins that block eukaryotic protein synthesis.

Southern blot - A technique used for searching for a specific DNA fragment in which DNA restriction fragments are transferred from an agarose gel to a nitrocellulose filter, where the denatured DNA is then hybridized to a radioactive probe (blotting).

REFERENCES

Abbaszadegan, M.; Huber, M.S.; Gerba, C.P.; and Pepper, I.L. Detection of viable *Giardia* cysts by amplification of heat shock induced mRNA. *Appl. Environ. Microbiol.* **1997**, *63*, 324-328.

Abbaszadegan, M.; DiGiovanni, G.; Czajka, J.; LeChevallier, M. Development of a PCR-Based Kit for Detection of *Cryptosporidium parvum* Using a Fluorescent Homogeneous Format. *Proceedings AWWA Water Quality Technology Conference, American Water Works Association*, Denver, 1998.

Al-Hendy, A.; Toivanen, P.; Skurnik, M. Expression cloning of *Yersinia enterocolitica* O:3 rfb gene cluster in *Escherichia coli* K12. *NCBI* **1991,** 10, 1: 47-59.

Araujo, R.M.; Arribas, R.M.; Pares, R. Distribution of *Aeromonas* species in water with different levels of pollution. *J. Appl. Bacteriol* **1991,** 71, 182-186.

Armstrong, D.; Gold, J.W.; Dryjanski, J.; Whimbey, E.; Polsky, B.; Hawkins, C.; Brown, A.E.; Bernard, E.; Kiehn, T.E. Treatment of infections in patients with the acquired immunodeficiency syndrome. *Ann Intern Med* **1985**, l03, 738-743.

Armon, R.; Payment, P. A modified m-CP medium for enumerating *Clostridium pefringens* from water samples. *Can. J. Microbiol* **1988**, 34, 78-79.

Aronson, T.W.; Holtzman, A.; Glover, N.; Boian, M.; Froman, S.; Berlin, O.G.W.; Hill, H.; Stelma, G., Jr. Comparison of large restriction Fragments of *Mycobacterium avium* isolates recovered from AIDS and non-AIDS patients with those of isolates from potable water. *J. Clinical Microbiol.* **1999,** 37, 4:1008-1012.

Ashbolt, N.J.; Kueh, C.S.; Grohmann, G.S. Significance of specific bacterial pathogens in the assessment of polluted receiving water of Sydney, Australia. *Water Sci. Technol.* **1993,** 27, 449-452

Bej, A.K.; DiCesare, J.L.; Haff, L.; Atlas, R.M. Detection of *Escherichia coli* and *Shigella* spp. in water by using the polymerase chain reaction and gene probes for *uid. Appl. Environ. Microbiol.* **1991**, 57, 1013-1017.

Bej, A.; Mahbubani, M.H.; Boyce, M.J.; Atlas, R.M. Detection of *Salmonella* spp. in oysters by PCR. *Appl. Environ. Microbiol.* **1994**, 60, 368-373.

Benaissa, M. Fine structure of viable coccoid forms of *Helicobacter pylori*: reversion to a dividing form. American Society for Microbiology, 95th General Meeting, 1995; Abstract 192.

Bhaduri, S.; Cottrell, B.; Pickard, A.R.Use of a single procedure for selective enrichment, isolation, and identification of plasmid-bearing virulent *Yersinia enterocolitica* of various serotypes from pork samples. *Appl. Environ. Micro.* **1997**, 63, 1657-1660.

Bissett, M.L.; Powers, C.; Abbott, S.L.; Janda, J.M. Epidemiologic investigations of *Yersinia enterocolitica* and related species: sources, frequency and serotype distribution. *J. Clin. Microbiol.* **1990**, 28, 910.

Bisson, J.W.; Cabelli, V.J. Membrane filter enumeration method for *Clostridium perfringens. Applied and Environmental Microbiology* **1979**, 37, 55-66.

Brown, P.K.; Romana, L.K.; Reeves, P.R. Molecular analysis of the *rfb* gene cluster of *Salmonella* serovar muenchen (strain M67): the genetic basis of the polymorphism between groups C2 and B. *Molecular Microbiology* **1992**, 6, 1385-1394.

Bruce, K.D.; Osborn, A.M.; Pearson, A.J.; Strike, P.; Ritchie, D.A. Genetic diversity within *mer* genes directly amplified from communities of non-cultivated soil and sediment bacteria. *Mol. Ecol.* **1995**, 4, 605-612.

Brunk, C.; Li, J. Quantitative Polymerase Chain Reaction Assay. In: *Environmental Molecular Microbiology: Protocols and Applications*; Rochelle, P.A.; Horizon Scientific Press, Wymondham, 2001; pp. 115-124.

Burke, V.; Robinson, J.; Gracey, M.; Peterson, D.; Partridge, K. Isolation of *Aeromonas hydrophila* from a metropolitan water supply: Seasonal correlation with clinical isolates. *Appl. Environ. Microbiol.* **1984**, 48, 361-366.

Burr, M.; Camper, A.K.; De Leon, R.; Hacker, P. Colonization of Biologically Active Filter Media with Pathogens. *American Water Works Association Research Foundation,* 2000.

Cabelli, V.J.; Dufour, A.P.; McCabe, L.J.; Levin, M.A. Swimming associated gastroenteritis and water quality. *Amer. J. Epidem.* **1982**, 115, 606-616.

Chapron, C.D.; Ballester, N.A.; Fontaine, J.H.; Frades, C.N.; Margolin, A.B. Detection of astroviruses, enteroviruses, and adenovirus types 40 and 41 in surface waters collected and evaluated by the information collection rule and an integrated cell culture-nested PCR procedure. *Appl. Environ. Micro.* **2000**, 66, 2520-2525.

Chauret, C.; Armstrong, N.; Fisher, J.; Sharma, R.; Springthorpe, S.; Sattar, S. Correlating *Cryptosporidium* and *Giardia* with microbial indicators. *J. Amer. Wat. Works Assoc.* **1995**, 87, 76-84.

Chen, W.; Martinez, G.; Mulchandani, A. Molecular beacons: A real-time polymerase chain reaction assay for detecting *Salmonella. Analytical Biochem.* 280:166-172

Cherry, W.B.; Hanks, J.B.; Thomason, B.M.; Murlin, A.M.; Biddle, J.W.; Croom, J.M. 1972. Salmonellae as an Index of Pollution of Surface Waters. *Appl. Microbiol.* **2000**, 24, 334-340.

Cody, R.M.; Tischer, R.G. Isolation and frequency of occurrence of *Salmonella* and *Shigella* in stabilization ponds. *JWPCF.* **1965**, 37, 1399.

Collins, J.M.; Yates, M.D. Mycobacteria in water. *J. of App. Bact.* **1984**, 57, 193-211

Colwell, R.R.; Grimes, D.J. Semantics and Strategies. In: *Nonculturable Microorganisms in the Environment*. Colwell, R.R., Grimes, D. J., Eds. ASM Press, Washington, D.C., 2000; pp.1-6.

Cooper, L.A.; Daum, L.T.; Roudabush, R.; Lohman, K.L. Molecular Identification of *Campylobacter coli* and *Campylobacter jejuni* Using Automated DNA Extraction and Real-Time PCR with Species-Specific TaqMan Probes. In: *Rapid Cycle Real-Time PCR- Methods and Applications*, Reischl, U., Wittwer, C., Cockerill, F., Eds. Springer-Verlag, Berlin, 2002; pp. 155-159.

Cover, T.L.; Blaser, M.J. *Helicobacter pylori*: A bacterial cause of gastritis, peptic ulcer disease, and Gastric Cancer. *ASM News*. **1995**, 61, 1:21-26.

Craun, G.F.; Berger, P.S.; Calderon, R.L. Coliform bacteria and waterborne disease outbreaks. *Jour. AWWA*. **1997**, 89, 96-104

Davis, E.M.; Casserly, D.M.; Moore, J.D. Bacterial relationships in stormwaters. *Water Resources Bulletin* **1977**, 13, 895-905.

Debartolomeis, J.; Cabelli, V.J. Evaluation of an *Escherichia coli* host strain for enumeration of F male-specific bacteriophages. *Appl. Environ. Microbiol.* **1991**, 57, 1301.

DeLong, E. F.; Franks, D.G.; Alldrege, A.L. Phylogenetic diversity of Aggregate-attached vs. fee-living marine bacterial assemblages. *Limnol. Oceanogr.* **1993**, 38, 924-934.

Desmarchelier, P.M.; Bilge, S.S.; Fagan, N.; Mills, L.; Vary, J.C.; Tarr, P.I. A PCR specific for *Escherichia coli* O157 based on the *rfb* locus encoding O157 lipopolysaccharide. *J. Clin. Microbiol.* 1998, 36, 1801-1804.

Dorsch, M.; Ashbolt, N.J.; Cox, P.T.; Goodman, A.E. Rapid identification of *Aeromonas* species using 16S rDNA targeted oligonucleotide primers: A molecular approach based on screening of environmental isolates. *J. Appl. Bacteriol.* **1994**, 77, 6:722-726.

Dowd, S.E.; Gerba, C.P.; Pepper, I.L. Confirmation of human pathogenic microsporidia *Enterocytozoon bieneusi, Encephalitozoon intestinalis* and *Vittaforma corneae* In water. *Appl. Environ. Microbiol.* **1998**, 64, 9:3332-3335.

Dutka, B.J.; Bell, J.B. Isolation of Salmonellae from moderately polluted waters. *J. Water Pollut. Control Fed.* **1973**, 45, 316-324

Enroth, H.; Engstrand, L. Immunomagnetic separation and PCR for detection of *Helicobacter pylori* in water and stool specimens. *J. Clinical Microbiology* **1995**, 33, 2162-2165.

Fach, P.; Popoff, M.R. Detection of enterotoxigenic *Clostridium perfringens in* food and fecal samples with a duplex PCR and the slide latex agglutination test. *Appl. Environ. Micro.* **1997**, 63, 4232-4236.

Fang, Y.; Wu, W.; Pepper, J.L.; Larsen, J.L.; Marras, S.A.E.; Nelson, E.A.; Epperson, W.B.; Christopher-Hennings, J. Comparison of real-time, quantitative PCR with molecular beacons to nested PCR and culture methods for detection of *Mycobacterium avium* subsp. *paratuberculosis* in bovine fecal samples. *J. Clinical Microbiol.* **2002**, 40, 287-291.

Feng, P. *Escherichia coli* serotype 0157:H7: novel vehicles of infection and emergence of phenotypic variants. *Emerging Infec. Dis*. **1995**, 2, 47.

Fenwick, S.G.; McCarty, M.D. *Yersinia enterocolitica* is a common cause of gastroenteritis in Auckland. *N.Z. Med. J.* **1995**, 108, 269.

Ferguson, C.M.; Coote, B.G.; Ashbolt, N.J.; Stevenson, I.M. Relationships between indicators, pathogens and water quality in an estuarine system. *Water Res.* **1996**, 30, 2045-2054.

Field, K.G.; Gordon, D.; Wright, T.; Rappe, M.; Urback, E.; Vergin, K.; Giovannoni, S.J. Diversity and depth-specific distribution of SAR11 cluster rRNA genes from marine planktonic bacteria. *Appl. Environ. Microbiol.* **1997**, 63, 63-70.

Fliermans, C.B.; Cherry, W.B.; Orrison, L.H.; Smith, S.J.; Tison, D.L.; Pope, D.H. Ecological Distribution of *Legionella pneumophila*. *Appl. Environ. Microbiol*. **1981**, 41, 9-16.

Fortin, N.Y.; Mulchandani; A.; Chen, W. Use of real-time polymerase chain reaction and molecular beacons for the detection of *Escherichia coli* O157:H7. *Analytical Biochem.* **2001**, 289: 281-288

Fuhrman, J.A.; McCallum, K.; Davis, A.A. Phylogenetic diversity of subsurface marine microbial communities from the Atlantic and Pacific Oceans. *Applied and Environmental Microbiology* **1993**, 59, 1294-1302.

Ginsburg, W.; Crawford, B.; Knipper, J.J. Filter-fluorescent antibody technique for rapid screening of indicator organisms. *J. Am. Wat. Works Assoc*. **1972**, 64, 499-505.

Giovannoni, S. The Polymerase Chain Reaction. In*: Nucleic Acid Techniques in Bacterial Systematics*. Eds: Stackebrandt, E., Goodfellow, M. John Wiley, Chichester, 1991; pp.177-203.

Glover, N.; Hilbourne, L.H.; Lee, D.; Blixt, H.C.; Liu, P. I. Disseminated *Mycobacterium avium* complex in patients with AIDS. *Laboratory Medicine*1995, 26, 204–209.

Glover, N.; Holtzman, A.; Aronson, T.; Froman, S.; Berlin, O.G.W.; Dominguez, P.; Kunkel, K.A.; Overturf, G.; Stelma, G. Jr.; Smith, C.; Yakrus, M. The isolation and identification of *Mycobacterium avium* complex (MAC) recovered from Los Angeles potable water, a possible source of infection in AIDS patients. *Intl. J. Environ. Health Res.* **1994,** 4, 63-72.

Hardy, W.D.; Feinberg, J.D.; Finkelstein, M.; Power, M.E.; He, W.; Kaczka, C.; Frame, P.T.; Holmes, M.; Waskin, H.; Fass, R.J.; Powderly, W.G.; Steigbigel, R.T.; Zuger, A.; Holzman, R.S. A control trial of trimethoprim-sulfamethoxazole or aerosolized pentamidine for secondary prophylaxis of *Pneumocystis carinii* pneumonia in patients with the acquired immunodeficiency syndrome. *New England Journal of Medicine* **1992**, 327, 1842-1848.

Havelaar, A.H.; van Olphen, M.; Drost, Y.C. F-Specific RNA Bacteriophages are Adequate Model Organisms for Enteric Viruses in Fresh Water. *Appl. Environ. Microbiol.* **1993**, 59, 2956-62.

Havelaar, A.H.; During, M.; Versteegh, J.F.M. Ampicillin-dextrin agar medium for the enumeration of *Aeromonas* species in water by membrane filtration. *Journal of Applied Microbiology* **1987**, 62, 279-287.

Horsburgh, C.R. Mycobacterium avium complex infection in the acquired immunodeficiency syndrome. *N. Engl. L. Med.* **1991**, 324, 1332-38.

Howgrave-Graham, A.R.; Steyn, P.L. Application of the fluorescent antibody technique for the detection of *Sphaerotilus natans* in activated sludge. *Appl. Environ. Microbiol.* **1988**, 54, 799-802.

Huang, P.W.; Laborde, D.; Land, V.R.; Matson, D.O.; Smith, A.W.; Jiang, X. Concentration and detection of calciviruses in water samples by reverse transcription-PCR. *Appl. Environ. Microbiol.* **2000**, 66, 4383-4388.

Hugenholtz, P.; Goebel, B.M. The Polymerase Chain Reaction as a Tool to Investigate Microbial Diversity in Environmental Samples. In*: Environmental Molecular Microbiology: Protocols and Applications*, P. A. Rochelle, Ed. Horizon Scientific Press, Wymondham, 2001; pp. 31-41.

Iwade, Y.; Yamauchi, A.; Sugiyama, A.; Nakayama, O.; Kumazawa, N.H. Quantification of Thermostable Direct Hemolysin-Producing *Vibrio parahaemolyticus* from Foods Assumed to Cause Food Poisoning Using the LightCycler Instrument. In*: Rapid Cycle Real-Time PCR-Methods and Applications*, Reischl, U.; Wittwer, C.; Cockerill, F., Eds., Springer-Verlag, Berlin, 2002; pp. 171-176.

Kapperud, G.; Vardund, T.; Skjerve, E.; Hornes, E.; Michaelsen,T.E. Detection of pathogenic *Yersinia enterocolitica* in foods and water by immunomagnetic separation, nested polymerase chain reactions, and colorimetric detection of amplified DNA. *Appl Environ. Microbiol.* **1993**, 59, 2938-2944.

Kinde H.; Adelson, M.; Ardans, A.; Little, E.H.; Willoughby, D.; Berchtold, D.; Read, D.H.;

Breitmeyer, R.; Kerr, D.; Tarbell, R.; Hughes, E. Prevalence of *Salmonella* in municipal sewage treatment plant effluents in southern California. *Avian Dis.* **1997**, 41, 392-398

Klein, P.D.; Graham, D.Y.; Gaillour, A.; Opekun, A.R.; Smith, E.O. Water source as risk factor for *H. pylori* infection in Peruvian children. *Lancet.* 1991, 337, 1503-1505.

Kott Y. Estimation of low numbers of *Escherichia coli* bacteriophage by use of the most probable number method. *Appl. Environ. Microbiol.* 1966, 14, 141.

Lage, A.P.; Godfroid, E.; Fauconnier, A.; Burette, A.; Butzler, J.; Bollen, A.; Glupczynski, Y. Table 2.4 Diagnosis of *Helicobacter pylori* Infection by PCR: Comparison with Other Invasive Techniques and Detection of *cag*A Gene in Gastric Biopsy Specimens. *J. Clinical Microbiol.* **1995**, 33, 2752-2756

Lanciott, R.S., and A.J. Kerst. 2001. Nucleic Acid Sequence-Based Amplification Assays for Rapid Detection of West Nile and St. Louis *Encephalitis* Viruses. *J. Clinical Microbiol.*39:4506-4513.

Lang, A.L.; Tsai, Y.L.; Mayer, C.L.; Patton, K.C.; Palmer, C. J. Multiplex PCR for detection of the heat-labile toxin gene and shiga-like toxin I and II genes in *Escherichia coli* isolated from natural waters. *Appl. Environ. Microbiol.* **1994**, 60, 3145-3149.

Levin, M.A. Quantitative large-volume sampling technique. *Appl. Environ. Microbiol.* **1974**, 28, 515.

Lindell, S.S.; Quinn, P. Shigella sonnei isolated from well water. *Appl. Environ. Microbiol.* **1973**, 26, 424.

Mager, A.L.; Standridge, J.; Kluender, S.M.; Peterson, L.L. Source and Occurrence of Pathogens in Watersheds. *Proceedings AWWA Water Quality Technology Conference*, American Water Works Association, 1998.

Makintubee, S.; Mallonee, J.; Istre, G. Shigellosis outbreak associated with swimming. *AJPH.* **1987**, 77, 166.

Marchesi, J.R. Primer Design for PCR Amplification of Environmental DNA Targets. In: *Environmental Molecular Microbiology: Protocols and Applications*, P. A. Rochelle, Ed., Horizon Scientific Press, Wymondham, 2001; pp. 43-54.

Maurer, J.J.; Schmidt, D.; Petrosko, P.; Sanchez, S.; Bolton, L.; Lee, M.D. Development of primers to O-antigen biosynthesis genes for specific detection of *Escherichia coli* O157 by PCR. *Appl. Environ. Microbiol.* **1999**, 65, 2954-2960.

Moriñigo, M.A.; Còrnax, R.; Muñoz, P.R. Relationships between *Salmonella spp* and indicator microorganisms in polluted natural waters. *Water Resource.* 1990, 24, 117.

Morinigo, M.A. Laboratory study of several enrichment broths for the detection of

Salmonella spp. particularly in relation to water samples. *Journal of Applied Bacteriology* **1993**, 74, 330.

Morona, R.; Matthews, M.; Morona, J.; Brown, M.H. Regions of the cloned *Vibrio cholera* rfb genes needed to determine Ogawa form of the O-antigen. *Molecular and General Genetics* **1990**, 224, 405-412.

Moyer, N.P. Clinical significance of *Aeromonas* species isolated from patients with diarrhea. *J. Clin. Microbiol.* **1987**, 25, 2044-2048.

Mullis, K.B.; Ferre, F.; Gibbs, R.A. *The Polymerase Chain Reaction.* Birkhauser, Boston, 1994.

Nogva, H.K.; Rudi, K.; Naterstad, K.; Holck, A.; Lillehaug, D. Application of 5'- nuclease PCR for quantitative detection of *Listeria monocytogenes* in pure cultures, water, skim milk, and unpasteurized whole milk. *Appl. Environ. Microbiol.* **2000**, 66, 4266-4271.

Novick, R.J.; Moreno-Cabral, C.E.; Stinson, E.B.; Oyer, P.E.; Starnes, V.A.; Hunt, S.A.; Shumway, N.E. *Nontuberculous mycobacterial* infections in heart transplant recipients: a seventeen-year experience. *J Heart Transplant.* **1990**, 9, 357-363.

Oberst, R.D.; Hays, M.P.; Bohra, L.K.; Phebus, R.K.; Yamashiro, C.T.; Paszko-Kolva, C.; Sargeant, J.M.; Gillespie, J.R. PCR-based amplification and detection of *Escherichia coli* O157:H7 Utilizing an internal fluorogenic probe and the 5' nuclease (TaqMan™) assay. *Appl. Environ. Microbiol.* **1998**, 64, 3389-3396.

O'Conner, D.R. *Report of the Walkerton Inquiry*, Ontario Ministry of the Attorney General, 2002.

Okrend, A.; Rose, B.; Matner, R. A screening method for the isolation of *E. coli* 0157:H7 from ground beef. *J. Food Prot.* **1990a**, 53,941-943.

Okrend, A.; Rose, B.; Matner, R. An improved screening method for the detection and isolation of *E. coli* 0157:H7 incorporating the 3M Petrifilm Test Kit-HEC for hemorrhagic Escherichia coli 0157:H7. *J. Food Prot.* **1990b**, 53, 936-940.

Olsvik, O.; Strockbine, N.A. PCR Detection of Heat-stable, Heat-Labile, and Shiga-Like Toxin Genes in *Escherichia coli.* In: *Diagnostic Molecular Microbiology: Principles and Applications*, Persing, D.H., Smith, T.F., Tenover, F.C., White, T.J. American Society for Microbiology, Washington, D.C., 1993; pp. 271-276.

Oyofo, B.A.; Rollins, D.M. Efficacy of filter types for detecting *Campylobacter jejuni* and *Campylobacter coli* in environmental water samples by polymerase chain reaction. *Appl. Environ. Microbiol.* **1993**, 59, 4090.

Pace, N.R. A molecular view of microbial diversity and the biosphere. *Science* **1997**, 276, 734-740.

Patra, G.; Vaissaire, J.; Weber-Levy, M.; Le Doujet, C.; Mock, M. Molecular characterization of *Bacillus* strains involved in outbreaks of anthrax in France in 1997.*Journal of Clinical Microbiology* **1998**, 36, 3412-3414.

Payment, P.; Gamache, F.; Paquette, G. Comparison of microbiological data from two water filtration plants and their distribution systems. *Water Sci. Technol.* 1989, 21, 287-289.

Payment, P.; Franco, E. *Clostridium perfringens* and somatic coliphages as indicators of the efficiency of drinking water treatment for viruses and protozoan cysts. *Appl. Environ. Microbiol.* **1993**, 59, 2418-2424.

Presnell, M.W. Use of the membrane filter and a filter aid for concentrating and enumerating indicator bacteria and *Salmonella* from estuarine waters. *Wat. Res.* **1976**, 10, 549.

Prince, D.S.; Peterson, D.D.; Steiner, R.M.; Gottlieb, J.E.; Scott, R.; Israel, H.L.; Figueroa, W.G.; Fish, J.E. Infections with *Mycobacterium avium* complex in patients without predisposing conditions. *N. Engl. J. Med.* **1989**, 321, 863-868.

Rappold, H.; Bolderdijk, R. Modified lysine iron agar for isolation of *Salmonella* from food. *Appl. Environ. Microbiol.* 1979, 38, 162-163.

Reasoner, D.J. Pathogens in Drinking Water-Are There Any New Ones." In *Proc. Water Qual. Technol. Conf.* Orlando, Fla. American Water Works Association, Denver, Co., 1991.

Rochelle, P.A.; Cragg, B.; Fry, J. C.; Parkes, R.J.; Weightman, A.J. Effect of sample handling on estimation of bacterial diversity in marine sediments by 16S rRNA gene sequence analysis. *FEMS Microbiol. Ecol.* **1994,** 15, 215-226.

Rochelle, P.A.; Will, J.A. K.; Fry, J.C.; Jenkins, G.J.S.; Parkes, R.J.; Weightman, A.J. Extraction and amplification of 16S rRNA genes from deep marine sediments and seawater to assess bacterial community diversity. In: *Nucleic Acids in the Environment*, Trevors, J.T.; van Elsas, J.D., Eds., Springer-Verlag, Heidelberg, 1995; pp. 219-239.

Rochelle, P.A. Environmental Molecular Microbiology: Introduction. In: *Environmental Molecular Microbiology: Protocols and Applications*, Rochelle, P.A., Ed., Horizon Scientific Press, Wymondham, 2001a; pp. 1-13.

Rochelle, P.A. Detection of Protozoa in Environmental Water Samples. In: *Environmental Molecular Microbiology: Protocols and Applications*, Rochelle, P.A., Ed., Horizon Scientific Press, Wymondham, 2001b; pp. 91-114.

Rochelle, P.A. Evaluation of Genotyping Techniques for *Crytoporidium parvum.* AWWA Research Foundation, Denver, 2001c.

Rosch, C.; Mergel, A.; Bothe, H. Biodiversity of denitrifying and dinitrogen-fixing bacteria in an acid forest soil. *Appl. Environ. Microbiol.* **2002**, 68, 3818-3829.

Rosenberg, M.L., K.K. Hazlet, J. Schaefer, J.G. Wells, and R.C. Pruneda. 1976. Shigellosis from swimming. *JAMA*. 236:1849.

Roszak, D.B.; Colwell, R.R. Survival strategies of bacteria in the natural environment. *Microbiological Reviews* **1987**, 51, 365-379.

Rudi, K.; Skulberg, O.M.; Larsen, F.; Jakobsen, K. S. Quantification of toxic cyanobacteria in water by use of competitive PCR followed by sequence-specific labeling of oligonucleotide probes. *Appl. Environ. Microbiol.* **1998,** 64, 2639-2643.

Sails, A.D.; Bolton, F.J.; Fox, A.J.; Wareing, D.R.A.; Greenway, D.L.A. Detection of *Campylobacter jejuni* and *Campylobacter coli* in environmental waters by PCR enzyme-linked immunosorbent assay. *Appl. Environ. Microbiol.* **2002,** 68, 1319-1324.

Schoolnik, G.K. PCR Detection of *Shigella* species and enteroinvasive *Escherichia coli*. In: *Diagnostic Molecular Microbiology: Principles and Applications,* Persing, D.H., Smith, T.F., Tenover, F.C., White, T.J., American Society for Microbiology, Washington, D.C., 1993; pp. 277-281.

Sharma, V.K.; Dean-Nystrom, E.A.; Casey, T.A. Semi-automated fluorogenic PCR Assays (TaqMan) for rapid detection of *Escherichia coli* O157:H7 and other Shiga toxigenic *E. coli*. *Mol. Cell. Probes* **1999,** 13, 291-302.

Sinigalliano, C.D.; Kuhn, D.N.; Jones, R.D. Amplification of the amo A gene from diverse species of ammonium-oxidizing bacteria and from an indigenous bacterial population from seawater. *Appl. Environ. Microbiol.* **1995**, 61, 2702-2706.

Sobsey , M.D.; Battigelli, D.A.; Handzel ,T.R.; Schwab, K.J. Male-Specific Coliphages as Indicators of Viral Contamination of Drinking Water. American Water Works Association Research Foundation, Denver, 1995.

Sommerville, C.C.; Knight, I.T.; Straub, W.L.; Colwell, R.R. Simple, rapid method for direct isolation of nucleic acids from aquatic environments. *Appl Environ Microbiol.* **1989**, 55, 548-554

Spinner, M.L.; Di Giovanni, G.D. Detection and identification of mammalian reoviruses in surface water by combined cell culture and reverse transcription-PCR. *Appl Environ. Microbiol.* **2001**, 67, 3016-3020.

Strockbine, N.A.; Wells, J.G.; Bopp, C.A.; Barrett, T.J., Overview of Detection and Subtyping Methods. In: *Escherichia coli O157:H7 and Other Shiga Toxin-Producing E. coli Strains*. Kaper, J.B., O'Brien, A.D., Eds. ASM Press, Washington, D.C., 1998. pp. 331-356.

Sturbaum, G.D.; Ortega, Y.R.; Gilman, R.H.; Sterling, C.R.; Cabrera, L.; Klein, D. A. Detection of *Cyclospora cayetanensis* in wastewater. *Appl. Environ. Microbiol.* **1998,** 64, 2284-2286.

Takewaki, S.; Okuzumi, K.; Ishiko, H.; Nakahara, K.; Ohkubo, A.; Nagai, R. Genus-specific polymerase chain reaction for the *Mycobacterial dnaJ* gene and species-specific oligonucleotide probes. *J. Clin. Microbiol.* **1993,** 31, 446-50.

Trevors, J.T.; van Elsas, J.D. *Nucleic Acids in the Environment*. Springer-Verlag, Berlin, 1995.

Tsai, Y.L.; Rochelle, P.A. Extraction of Nucleic Acids from Environmental Samples. In: *Environmental Molecular Microbiology: Protocols and Applications*, Rochelle, P.A., Ed., Horizon Scientific Press, Wymondham, 2001; pp. 15-30.

Tyagi, S.; Kramer, F.R. Molecular beacons: Probes that fluoresce upon hybridization. *Nature Biotechnology* **1996,** 14, 303-308.

Van Sluis, R.J.; Yanko, W.A. The Fate of *Salmonella* sp. Following Dechlorination of Tertiary Sewage Effluents. Project report submitted in partial fulfillment of U.S. EPA Contract No. 14-12-150, Dechlorination of Wastewater: State-of-the-art Field Survey and Pilot Plant Studies, 1977.

Vial, P.A.; Robbins-Browne, R.; Lior, H.; Prado, V.; Kaper, J.B.; Nataro, J.B., Meneval, D.; Elsayed, A.-E.-D.; M.M. Levine. Characterization of enteroadherent-aggregative Escherichia coli, a putative agent of diarrheal disease. *J. Infect. Dis.* **1988,** 158, 70.

Wang, L.; Reeves, P.R. Organization of *Escherichia coli* O157 O antigen gene cluster and identification of its specific genes. *Infection and Immunity* **1998,** 66:3545-3551.

Weagant, S.D.; Kaysner, C.A. Modified enrichment broth for isolation of *Yersinia enterocolitica* from nonfood sources. *Appl. Environ. Microbiol.* **1983**, 45, 468-471.

Wellinghausen, N.; Frost, C.; Marre, R. Detection of *Legionellae* in hospital water samples by quantitative real-time LightCycler PCR. *Appl. Environ. Microbiol.* **2001,** 67, 3985-3993.

West, A.P.; Millar, M.R.; Tompkins, D.S. Survival of *Helicobacter pylori* in water and saline. *J. Clin. Pathol.* **1990,** 43, 609.

Wolk, D.M.; Schneider, S.K.; Wengenack, N.L.; Sloan, L.M.; Rosenblatt, J.E. Real-time PCR method for detection of *Encephalitozoon intestinalis* from stool specimens. *J. Clinical Microbiol.* **2002,** 40, 3922-3928.

Yanko, W.A.; Jackson, J.L.; Williams, F.P.; Walker, A.S.; Castillo, M.S. An unexpected temporal pattern of coliphage isolation in groundwaters sampled from wells at varied distances from reclaimed water recharge sites. *Water Research.* **1999,** 33, 1.

Xiao, L.; Alderisio, K.; Limor, J.; Royer, M.; Lal, A. Identification of species and sources of *Cryptosporidium* oocysts in storm waters with a small-subunit rRNA-based diagnostic and genotyping tool. *Appl. Environ. Microbiol.* **2000,** 66, 5492-5498.

Xu, H.H.; Tabia, F.R. Ribulose-1,5-bisphosphate carboxylase/oxygenase gene expression and diversity of Lake Erie planktonic microorganisms. *Appl. Environ. Microbiol.* **1996,** 62, 1913-1921.

CONTACT INFORMATION

This was a collaborative study conducted by the County Sanitation Districts of Los Angeles County (CSDLAC), the Metropolitan Water District of Southern California (MWDSC), and the University of California, Riverside (UCR). Each project participant had primary project responsibilities as follows:

CSDLAC	Large volume (GF/F-DE) concentration method evaluation and validation, and all culture-based analyses.
MWDSC	Molecular methods and analyses
UCR	Molecular beacon evaluation

William A. Yanko is currently an environmental microbiology consultant. He retired from CSDLAC prior to completion of the project but continued as the lead P.I. through completion of the project report. He can be contacted by e-mail at byanko@mac.com. CSDLAC agency contacts concerning this project are Shawn Thompson, Ph.D., Microbiology Laboratory Supervisor (sthompson@lacsd.org), or Alan Walker, Senior Microbiologist (awalker@lacsd.org). Both can be contacted at San Jose Creek Water Quality Laboratory, 1965 Workman Mill Road, Whittier, CA 90601. Tel: 562-699-7411.

Ricardo De Leon, Ph.D. (Microbiology Unit Manager, 909-392-5155) and Paul A. Rochelle, Ph.D. (Principal Microbiologist, 909-392-5155) can be contacted at: The Metropolitan Water District of Southern California, Water Quality Laboratory, 700 Moreno Avenue, La Verne, CA 91750 or via e-mail (ricardo_deleon@mwdh2o.com or prochelle@mwdh2o.com).

Wilfred Chen, Ph.D. Associate Professor can be contacted at: Department of Chemical and Environmental Engineering, University of California, Riverside, CA 92521, Tel: 909-787-2473, or via e-mail (wilfred@engr.ucr.edu).

WASTEWATER UTILITY

Alabama
Montgomery Water Works & Sanitary Sewer Board

Alaska
Anchorage Water & Wastewater Utility

Arizona
Gila Resources
Glendale, City of, Utilities Department
Mesa, City of
Peoria, City of
Phoenix Water Services Department
Pima County Wastewater Management

Arkansas
Little Rock Wastewater Utility

California
Calaveras County Water District
Central Contra Costa Sanitary District
Contra Costa Water District
Crestline Sanitation District
Delta Diablo Sanitation District
Dublin San Ramon Services District
East Bay Dischargers Authority
East Bay Municipal Utility District
El Dorado Irrigation District
Fairfield-Suisun Sewer District
Fresno Department of Public Utilities
Irvine Ranch Water District
Las Virgenes Municipal Water District
Livermore, City of
Lodi, City of
Los Angeles, City of
Los Angeles County, Sanitation Districts of
Napa Sanitation District
Orange County Sanitation District
Palo Alto, City of
Riverside, City of
Sacramento Regional County Sanitation District
San Diego Metropolitan Wastewater Department, City of
San Francisco, City & County of
San Jose, City of
Santa Barbara, City of
Santa Rosa, City of
South Bayside System Authority
South Coast Water District
South Orange County Wastewater Authority
Union Sanitary District
West Valley Sanitation District

Colorado
Boulder, City of
Colorado Springs Utilities
Littleton/Englewood Water Pollution Control Plant
Metro Wastewater Reclamation District, Denver

Connecticut
The Mattabasset District
New Haven, City of, WPCA

District of Columbia
District of Columbia Water & Sewer Authority

Florida
Broward, County of
Fort Lauderdale, City of
Gainesville Regional Utilities
JEA
Miami-Dade Water & Sewer Authority
Orange County Utilities Department
Orlando, City of
Reedy Creek Improvement District
Seminole County Environmental Services
St. Petersburg, City of
Stuart Public Utilities
Tallahassee, City of
Tampa, City of
Toho Water Authority
West Palm Beach, City of

Georgia
Atlanta Department of Watershed Management
Augusta, City of
Clayton County Water Authority
Cobb County Water System
Columbus Water Works
Fulton County
Gwinnett County Department of Public Utilities
Macon Water Authority
Savannah, City of

Hawaii
Honolulu, City and County of

Illinois
American Bottoms Wastewater Treatment Plant
Greater Peoria Sanitary District
Kankakee River Metropolitan Agency
Metropolitan Water Reclamation District of Greater Chicago
North Shore Sanitary District
Wheaton Sanitary District

Indiana
Fort Wayne, City of

Iowa
Ames, City of
Cedar Rapids Wastewater Facility
Des Moines Metro Wastewater Reclamation Authority
Iowa City, City of

Kansas
Johnson County Unified Wastewater Districts
Unified Government of Wyandotte County/Kansas City, City of

Kentucky
Louisville & Jefferson County Metropolitan Sewer District

Louisiana
Sewerage & Water Board of New Orleans

Maine
Bangor, City of
Portland Water District

Maryland
Anne Arundel County Bureau of Utility Operations
Howard County Department of Public Works
Washington Suburban Sanitary Commission

Massachusetts
Boston Water & Sewer Commission
Upper Blackstone Water Pollution Abatement District

Michigan
Ann Arbor, City of
Detroit, City of
Holland Board of Public Works
Lansing, City of
Owosso Mid-Shiawassee County WWTP
Saginaw, City of
Wayne County Department of Environment
Wyoming, City of

Minnesota
Rochester, City of
Western Lake Superior Sanitary District

Missouri
Independence, City of
Kansas City Missouri Water Services Department
Little Blue Valley Sewer District

Nebraska
Lincoln Wastewater System

Nevada
Henderson, City of

New Jersey
Passaic Valley Sewerage Commissioners

New York
New York City Department of Environmental Protection
Rockland County Solid Waste Management Authority/ Sewer District

North Carolina
Cary, Town of
Charlotte/Mecklenburg Utilities
Durham, City of
Metropolitan Sewerage District of Buncombe County
Orange Water & Sewer Authority

Ohio
Akron, City of
Butler County Department of Environmental Services
Columbus, City of
Metropolitan Sewer District of Greater Cincinnati
Northeast Ohio Regional Sewer District
Summit, County of

Oklahoma
Tulsa, City of

Oregon
Clean Water Services
Eugene/Springfield Water Pollution Control
Water Environment Services

Pennsylvania
Philadelphia, City of
University Area Joint Authority, State College

South Carolina
Charleston Commissioners of Public Works
Mount Pleasant Waterworks & Sewer Commission
Spartanburg Sanitary Sewer District

Tennessee
Cleveland, City of
Knoxville Utilities Board
Murfreesboro Water & Sewer Department
Nashville Metro Water Services

Texas
Austin, City of
Dallas Water Utilities
Denton, City of
El Paso Water Utilities
Fort Worth, City of
Gulf Coast Waste Disposal Authority

WERF SUBSCRIBERS

Houston, City of
San Antonio Water System
Trinity River Authority

Utah
Salt Lake City Corporation

Virginia
Alexandria Sanitation Authority
Arlington, County of
Fairfax County Virginia
Hampton Roads Sanitation District
Henrico, County of
Hopewell Regional Wastewater Treatment Facility
Loudoun County Sanitation Authority
Lynchburg Regional WWTP
Prince William County Service Authority
Richmond, City of
Rivanna Water & Sewer Authority

Washington
Edmonds, City of
Everett, City of
King County Department of Natural Resources
Seattle Public Utilities
Sunnyside, Port of
Yakima, City of

Wisconsin
Green Bay Metro Sewerage District
Madison Metropolitan Sewerage District
Milwaukee Metropolitan Sewerage District
Racine, City of
Sheboygan Regional Wastewater Treatment
Wausau Water Works

Australia
South Australian Water Corp.
Sydney Water Corp.
Water Corp. of Western Australia

Canada
Greater Vancouver Regional District
Toronto, City of, Ontario
Winnipeg, City of, Manitoba

Mexico
Servicios de Agua y Drenaje de Monterrey, I.P.D.

New Zealand
Watercare Services Limited

Singapore
Singapore Public Utilities Board

United Kingdom
Yorkshire Water Services Limited

STORMWATER UTILITY

California
Los Angeles, City of, Department of Public Works
Monterey, City of
Santa Rosa, City of

Colorado
Boulder, City of

Georgia
Griffin, City of

Iowa
Cedar Rapids Wastewater Facility
Des Moines Metro Wastewater Reclamation Authority

Kansas
Overland Park, City of

Kentucky
Louisville & Jefferson County Metropolitan Sewer District

Maine
Portland Water District

Minnesota
Western Lake Superior Sanitary District

North Carolina
Cary, Town of
Charlotte, City of, Stormwater Services

Pennsylvania
Philadelphia, City of

Tennessee
Chattanooga Storm Water Management

Washington
Seattle Public Utilities

STATE

Arkansas Department of Environmental Quality
Fresno Metropolitan Flood Control District, Calif.
Urban Drainage & Flood Control District, Colo.

CORPORATE

ADS Environmental Services
The ADVENT Group Inc.
Alan Plummer & Associates
Alden Research Laboratory
Alpine Technology Inc.
Aqua-Aerobic Systems, Inc.
Aquateam–Norwegian Water Technology Centre A/S
BaySaver Inc.
BioVir Laboratories, Inc.
Black & Veatch
Boyle Engineering Corporation
BPR CSO
Brown & Caldwell
Burns & McDonnell
CABE Associates Inc.
The Cadmus Group
Camp Dresser & McKee Inc.
Carollo Engineers Inc.
Carpenter Environmental Associates Inc.
CDS Technologies Inc.
Chemtrac Systems Inc.
CH2M HILL
Damon S. Williams Associates, LLC
David L. Sheridan, P.C.
Earth Tech Inc.
Eco-Matrix
Environmental Engineers International
EMA Inc.
The Eshelman Company Inc.
Finkbeiner, Pettis, & Strout Inc.
Freese & Nichols, Inc.
Frontier Geosciences Inc.
ftn Associates Inc.
Fuss & O'Neill, Inc.
Gannett Fleming Inc.
Golder Associates Inc.
Greeley and Hansen LLC
Hazen & Sawyer, P.C.
HDR Engineering Inc.
HNTB Corporation
HydroQual Inc.
Ingersoll-Rand Energy Systems
Insituform Technologies Inc.
Institute for Environmental Technology & Industry, Korea
Jacobson Helgoth Consultants Inc.
Jason Consultants Inc.
Jordan, Jones, & Goulding Inc.
KCI Technologies Inc.
Kelly & Weaver, P.C.
Kennedy/Jenks Consultants
Komline Sanderson Engineering Corporation
Lawler, Matusky & Skelly Engineers, LLP
Limno-Tech Inc.
Lombardo Associates Inc.
Malcolm Pirnie Inc.
McKim & Creed
MEC Analytical Systems Inc.
Metcalf & Eddy Inc.
MWH
New England Organics
Odor & Corrosion Technology Consultants Inc. (OCTC)
ONDEO Degremont Inc.
Oswald Green, LLC
PA Government Services Inc.
Parametrix Inc.
Parsons
Post, Buckley, Schuh & Jernigan
R&D Engineering/Conestoga Rover & Associates
The RETEC Group
RMC, Inc. (Raines, Melton & Carella)
R.M. Towill Corporation
Ross & Associates Ltd.
Royce Technologies
SAIC Maritime Technical Group
Sear-Brown
Short Elliott Hendrickson, Inc.
Stormwater Management, Inc.
Synagro Technologies, Inc.
Tetra Tech Inc.
Trojan Technologies Inc.
URS Corporation
Veolia Water NATC
Wade-Trim Inc.
Weston Solutions Inc.
Woodard & Curran
WRc/D&B, LLC
WWETCO, LLC
Zoeller Pump Company

INDUSTRY

American Electric Power
ChevronTexaco Energy Research & Technology Company
The Coca-Cola Company
Dow Chemical Company
DuPont Company
Eastman Kodak Company
Merck & Company Inc.
ONDEO Services
Procter & Gamble Company
PSEG Services Corp.
Severn Trent Services Inc.
Shell Global Solutions
Thames Water Plc
United Water Services LLC

Note: List as of 1/31/04

WERF Product Order Form

As a benefit of joining the Water Environment Research Foundation, subscribers are entitled to receive one complimentary copy of all reports and other products.Additional copies are available at cost (usually $10).To order your complimentary copy of a report, please w "free" in the unit price column.WERF keeps track of all orders.If the charge differs from what is shown here, we will call to confirm the before processing.

Name ________________ Title ________________

Organization ________________

Address ________________

City ________ State ________ Zip Code ________ Country ________

Phone ________ Fax ________ Email ________

Stock #	Product	Quantity	Unit Price	Total
			Postage & Handling	
			VA Residents Add 4.5% Sales Tax	
			Canadian Residents Add 7% GST	
			TOTAL	

Method of Payment: (All orders must be prepaid.)

❑ Check or Money Order Enclosed

❑ Visa ❑ Mastercard ❑ American Express

Account No. ________ Exp. Date ________

Signature ________

Shipping & Handling:

Amount of Order	United States	Canada & Mexico	All Others
Up to but not more than:	Add:	Add:	Add:
$20.00	$5.00*	$8.00	50% of amount
30.00	5.50	8.00	40% of amount
40.00	6.00	8.00	
50.00	6.50	14.00	
60.00	7.00	14.00	
80.00	8.00	14.00	
100.00	10.00	21.00	
150.00	12.50	28.00	
200.00	15.00	35.00	
More than $200.00	Add 20% of order	Add 20% of order	

*minimum amount for all orders

Note: Please make checks payable to the Water Environment Research Foundation.

To Order (Subscribers Only):

Log on to www.werf.org and clic on the "Product Catalog."

Phone: (703) 684-2470
Fax: (703) 299-0742.

WERF
Attn: Subscriber Services
635 Slaters Lane
Alexandria, VA 22314-1177

To Order (Non-Subscribers):

Non-subscribers may be able to orde WERF publications either through WEF (www.wef.org) or IWAP (www.iwapublishing.com).Visit WER website at www.werf.org for details.

IWA Publishing's authorised EU representative for General Product Safety Regulations is Diane D'Arras, 15 rue Duret, 75116 Paris, France, e-mail: safety@iwap.co.uk.

Printed and bound by CPI Group (UK) Ltd, Croydon, CR0 4YY
12/05/2026
02108893-0012